UNLIMITED LOVE

Unlimited Love

Altruism, Compassion, and Service

Stephen G. Post

Templeton Foundation Press
PHILADELPHIA AND LONDON

Templeton Foundation Press
Five Radnor Corporate Center, Suite 120
100 Matsonford Road
Radnor, Pennsylvania 19087
www.templetonpress.org

Designed and typeset by Gopa & Ted2
Printed by Versa Press, Inc.

Library of Congress Cataloging-in-Publication Data

Post, Stephen Garrard, 1951-
  Unlimited love : altruism, compassion, and service /
Stephen G. Post.
      p. cm.
Includes bibliographical references and index.
  ISBN 1-932031-31-6 (pbk. : alk. paper)
  1. Altruism. 2. Helping behavior. 3. Love. I. Title.

  BF637.H4 P67 2003
  177'.7—dc21   2002153955

Printed in the United States of America

03 04 05 06 07   10 9 8 7 6 5 4 3 2 1

CONTENTS

PREFACE

Sooner or later, all the peoples of the world will have to discover a way to live together in peace, and thereby transform this pending cosmic elegy into a creative psalm of brotherhood. If this is to be achieved, man must evolve for all human conflict a method which rejects revenge, aggression, and retaliation. The foundation of such a method is love.

THE REV. DR. MARTIN LUTHER KING, JR.

THE INSTITUTE for Research on Unlimited Love, a nonprofit organization, began in 2001 with initial funding from the John Templeton Foundation. Every institute owes the public a readable articulation of its subject matter. This Institute in particular requires such a statement because the word *love* is used in so many different ways, and yet our interest is specifically focused on the scientific understanding and practice of such remarkable phenomena as altruism, compassion, and service. This book is an initial effort to convey the meaning of generous unselfish love at the interface of science, ethics, and religion.

The Institute offers the following definition of unlimited love:

> The essence of love is to affectively affirm as well as to unselfishly delight in the well-being of others, and to engage in acts of care and service on their behalf; unlimited love extends this love to all others without exception, in an enduring and constant way. Widely considered the highest form of virtue, unlimited love is often deemed a Creative Presence underlying and integral to all of reality: participation in unlimited love constitutes the fullest experience of spirituality. Unlimited love may result in new relationships, and deep community may emerge around helping behavior, but this is secondary. Even if connections and relations do not emerge, love endures.

The mission of the Institute for Research on Unlimited Love is to help people to better understand their capacities for participation in unlimited love as the ultimate purpose of their lives. The Institute's goals are to:

(1) fund high-level scientific research on altruistic and unlimited love;

(2) develop a sustained dialogue between religion and science on the meaning and significance of unlimited love through publications and conferences;

(3) disseminate the real story of unlimited love as it is manifested in the helping behaviors of those whose lives are devoted to giving to others;

(4) enhance the practical manifestations of unlimited love across the full spectrum of human experience, including family life, education, leadership, community service, religion, and the professions, by providing conference opportunities and awards for innovative scholars and practitioners.

Our initial funded projects focus on six main program areas:

✦ *Human Development*—focusing on the biological, psychological, sociological, spiritual, and religious aspects of the human developmental trajectory;

✦ *Public Health and Medicine*—delving into questions of the effect of selfless love on mortality, therapy, and well-being;

✦ *Defining Mechanisms by Which Altruistic Love Affects Health*—exploring its effect on disease, immune cell function, learning, memory, and other health factors;

✦ *Other-Regarding Virtues*—examining the relationships among religion, spirituality, and altruism, and how they relate to social and interpersonal outcomes;

✦ *Evolutionary Perspectives on Other-Regard*—looking for possible connections between other-regarding love and the processes of natural selection;

✦ *Sociological Study of Faith-based Communities and their Activities in Relation to the Spiritual Ideal of Unlimited Love*—seeking, for example, to understand the nature of volunteerism and the ability of religiously motivated workers to combat antisocial behavior.

Research submissions to the Institute have covered a wide range of topics: deep altruism beyond kin boundaries; the physician-patient encounter; religion, giving, and volunteerism; generosity in organ donors and recipients; the healing benefits of love in veterans with Post Traumatic Stress Disorder; religious teachings and experiences in relation to loving behavior; uncompensated helping behavior in dolphins; the cognitive neuroscience of empathy; the behavioral neurology of love; putting love to the test in alcohol abuse rehabilitation; altruistic love in the family and early child development; and how faith-based communities develop expressions and understandings of altruistic caregiving.

The Institute has initiated collaborations with major national foundations and institutions to explore the place of unlimited love in the narrative of religious traditions, the humanities and arts, theology, ethics, education at all levels, and leadership. It extends its findings through conferences, publishing, national essay competitions for young people to express the importance of compassionate love in their development, and opportunities for scholars in science and religion to develop book proposals that will be supported after competitive review. The Institute's first major book, *Altruism and Altruistic Love: Science, Philosophy and Religion in Dialogue*, developed in collaboration with the Fetzer Institute, is available from Oxford University Press (2002). White papers and other high-level research papers can be found on the Institute's website at www.unlimitedloveinstitute.org.

Starting a research institute on a topic such as unlimited love is a bit of an adventure, and there was no clear sense of what the response might be. But receiving more than three hundred twenty Letters of Intent from leading researchers in response to a Request for Proposals disseminated in January of 2002 showed how much serious scientific interest there is in this topic. After only one year of existence, the Institute has already shown itself to be rapidly attaining its footing. Its advisory board includes Steven C. Rockefeller, Jr., a managing director of Deutsche Bank; George E. Vaillant, M.D., a professor of psychiatry at Harvard Medical School; Audrey R.

Chapman, director of the Science and Human Rights Program and the Program of Dialogue Between Science and Religion for the American Association for the Advancement of Science; Seyyed Hossein Nasr, Ph.D., professor of Islamic Studies at the George Washington University; and Dame Cicely Saunders, founder of the modern hospice movement. Former First Lady Rosalynn Carter, vice-chair of the board of trustees of the Carter Center in Atlanta and another member of the advisory board, recently wrote, "You are off to a great start in this important area of research, and you and your colleagues are to be commended."

The dominant paradigm of our time is scientific, and thus dialogues between science, religion, and ethics are vital to intellectual and practical progress. The scientific study of how ordinary good people give other-regarding love is important. What are their characteristics? What is their course of human development? Do they live relatively fulfilled, happy, healthy, and long lives in comparison with egoists? How did their parents influence them, if at all? As one of the twentieth century's greatest scientists of altruism, Pitirim Sorokin, wrote about the study of the "ways and powers of love":

> We have studied the negative types of human beings sufficiently—the criminal, the insane, the sinning, . . . and the selfish. But we have neglected the investigation of positive types of *Homo sapiens*—the creative genius, the saint, the "good neighbor." We know a great deal about the general characteristics of the subsocial types. But we know precious little of the general or typical properties of creative persons. What, if any, are the typical characteristics of altruistic persons?[1]

We need to better understand how lives of generous love come into being, and armed with such understanding, we should encourage the pedagogy of love.

How can science help? It is unlikely that science can fully explain the behavior of people who live in ways more or less consistent with unlimited love. Perhaps the most important thing we can do is simply to tell the stories of unlimited love as these brighten the world in which we live. Love is less taught didactically or studied scientifically than it is *transmitted* through models.[2]

Yet science holds out great potential in this field. Teilhard de Chardin commented that the scientific understanding of the power of unselfish love would be as significant in human history as the discovery of fire. Unselfish universal love is so important to our human future that we must examine it scientifically, thereby removing it from the domain of subjective "soft" truth and elevating it into that of objective "hard" truth. Why do we find more than one hundred thousand published scientific studies on depression and schizophrenia, and no more than a few dozen good studies on unselfish love? Because science has been largely focused on human deficits rather than on the positive side of our nature. Over the past fifteen years, my medical students have taken various forms of neurosis, psychosis, and personality disorder seriously since these are studied by scientific methods; yet they too often dismiss compassionate love for patients as "touchy-feely" because there is too little scientific attention given to it. The message is that compassion is not important enough to study. Why not bring these same scientific methods to the study of unselfish love? Why not apply every known scientific technique to the study of unselfish love, just as these have been widely applied to disease? What if we could absolutely prove that love heals mental illness and is vital to successful therapeutic outcomes in all areas of health care? What if we could absolutely prove that people who live more for others than for self have greater psychological well-being?

To some extent, we can study the opposites of love to understand what love requires. For example, Paul Connolly surveyed 352 children between three and six years of age from across Northern Ireland. Through the influence of the family, the local community, and the school, Roman Catholic and Protestant children have learned to loathe and fear one another even at these very young ages, deeply absorbing hatred and prejudice by age five. This early inculcation of hatred, reported in 2002, mirrors studies of the attitudes of Israeli and Palestinian children.[3] Yet such inculcation of hatred does not imply that loving-kindness is not within the repertoire of human nature. Even in William Golding's pessimistic novel *The Lord of the Flies,* a parable on human nature in which a group of English schoolboys is plane-wrecked on a deserted island and succumb to every form of selfishness and violence, young Simon remains kind and good to the end. We need to cultivate this goodness, and be certain that our religions and communities tap into this side of human nature rather than its opposite.

When science does focus on unselfish love, the media appear ready to take notice. A recent issue of the journal *Neuron* reports an Emory

University study revealing a biological basis for human cooperation.[4] Functional MRI scans have identified a "biologically embedded" basis for altruistic behavior, with several characteristic regions of the brain being activated when players of a game called "Prisoner's Dilemma" decide to trust each other and cooperate, rather than betray each other for immediate gain. This is, of course, a study of "I help you and then you will be inclined to help me," not of "I help others, and others gain not me." What might this study have to do with our interest in the latter form of love, which is not predicated on reciprocal response, and which makes no bargains and does not keep track of who reciprocates and who does not? It is perhaps the case that some of the same areas of the brain are involved in this purer form of love, or that such love is a transposition of cooperation to a higher level in which the agent perceives cooperation with God. We simply do not know, but will learn more. Do the great lovers of all humanity use the same regions of the brain highlighted in the Emory study, although freed from reciprocal considerations? Only further studies can answer this question. And are we so hard-wired for narrow loyalties and nationalistic "groupishness" that we can cooperate within our own group but not terribly well outside of it? In a time of escalated global conflict and weapons of mass destruction, learning more about how to encourage love for all humanity without exception is imperative.

In the final analysis, unlimited love is what God has for each and every one of us, and this is good news. The Institute seeks integral knowledge at the interface of science, human experience, and the underlying metaphysics of divine love.

Unlimited Love

INTRODUCTION

PROGRESS THROUGH LOVE

U NSELFISH LOVE for all people without exception is the most important point of convergence among all significant spiritualities and
religions. We marvel at the ways and power of love and find in it the best
hope for a far better human future. Innumerable everyday people excel in
loving-kindness, not just for their nearest and dearest, but as volunteers on
behalf of the neediest. Some people achieve miracles and become exemplars of generous pure love. How do our complex brains, unique imaginations, communicative abilities, reasoning powers, moral sense, and spiritual
promptings give rise to the remarkable and not at all uncommon practice
of unselfish love for our neighbors, or for those we do not even know? If
we could answer this question and harness the power of love, the world
would erupt into hope.

The question of how this unselfish love came to be led me to biology
and evolution as a college student. How, I wondered, could this generative
planet Earth give rise to a creature capable at its best of such remarkable
love? It led me to an interest in various eastern religions in my early twenties,
and eventually to the University of Chicago Divinity School, where in 1983
I wrote a doctoral dissertation on the idea of love within the context of
Christian thought. After several years of college teaching, I concluded that
the ideal of love had to be integrated with solid science in order to convince
students of the time-honored yet paradoxical truth that in the giving of
self lies the unsought discovery of self. I also became convinced that students only effectively study unselfish love when they practice it routinely,
and that teachers can only be effective when they are themselves practitioners. In 1988 I left a college in New York for the School of Medicine at
Case Western Reserve University, where for many years I have worked with
and studied individuals and families grappling to maintain the spirit of love

in the context of Alzheimer's disease. Unlimited love must bring resurrec-
tion-of-a-sort even to the most deeply forgetful, or so I hypothesized.

In 2001 my integrative journey around the theme of love took an unex-
pected turn. Sir John Templeton, who has devoted his resources to the
study of science and religion through the formation of the Templeton
Foundation, invited me to direct a new research initiative into what he
boldly named "unlimited love." This responsibility allowed me to facilitate the development of a positive scientific program integrating practice
with high-level empirical research, religious-ethical ideas, and metaphysics.
It was as though some creative higher presence in the universe had reached
into my life with a huge challenge: "If giving is the best way of living,
prove it!"

I grew to deeply appreciate Sir John's language of "unlimited love."
After all, what could unselfish love for every person without exception be
if not unlimited, at least in scope? Such love might sometimes have to be
constructively "tough" to be effective, and it might take many forms ranging from compassion to correction, but underlying all the expressions of
love is an affirmation of the goodness and potential in every life. Sir John
understood "unlimited love" at its highest as God's love for humanity, and
so did I. Ultimately, only divine love knows no limits. We human beings are
limited creatures indeed. But to the extent that we can and do achieve significant other-regarding love by our evolved human nature, or by participation in the divine through grace, we move progressively forward.

Most of us have encountered memorably unselfish, genuinely kind, and
deeply generous individuals, some of whom may have put themselves at
considerable risk in the service of perfect strangers. We are struck by the
emotional tone, intensity, and helping behavior of good parents, good
neighbors, good friends, and good servants of all imperiled and needy
people. It is natural to love one's children and friends, but less so to love
strangers, enemies, or those made unattractive by severe illness. Yet only in
leaning outward to all humanity does one transmit to children and friends
the higher purpose that can elevate their lives beyond the confines of near-
sighted emotional and material overindulgence. *Generous love for all others is
the main purpose of our life, the most enduring source of meaning and dignity, and the
basis for all lasting self-esteem.* This love seeks the good of the loved one and
leaves all the secondary effects, such as psychological well-being and elevated self-esteem, to take care of themselves.

The alternatives to growth in love include a drifting malaise devoid of purpose and accomplishment ("just hanging out"), the pursuit of ultimately meaningless selfish goals connected with narrow ambitions and materialism, or descent into hatred. Before September 11, 2001, there was April 20, 1999, when thirteen students were gunned down at Columbine High School in Colorado. We are astounded at the downward spiral of relatively young people—of whatever creed or nationality—into a negative vortex of hatred and murderous suicide. In the absence of adult mentors who encourage them in the ways of unselfish love, the influences on them range from distorted religious teachings to graphic media violence. A few weeks after 9/11, Fred Rogers, a Presbyterian minister who has devoted his life to the ideal of *agape* or unselfish Christian love, and who is now known by television reruns of "Mister Rogers' Neighborhood," was widely quoted in the media for his response to a reporter's question: "Mr. Rogers, what should parents tell their children?" Mr. Rogers was pensive, and then he answered what I believe we must all answer: "Tell them to keep their eyes on the helpers." By setting the eyes and intellects of the young on unselfish love, we encourage spiritual growth in ourselves and in them.

We need to better understand how love can penetrate and transform young and old from emptiness to fullness. We need to have faith in love despite the turbulence of our times and of all times, and we need to emerge into a world where we not only respect but actually cherish one another. We need to bring every scientific, educational, spiritual, ethical, and religious insight to this emergence. We need an evidence-based shift in our images of human motivational structure and a renewed confidence in the genuineness of our helping inclinations. We owe this to humanity, to our dignity, and to the future.

If I were forced to select a motto for the Institute for Research on Unlimited Love, it would be this: *In the giving of self lies the unsought discovery of self.* This fundamental law of life is simple, intuitive, and yet not clearly acknowledged. In essence, the paradoxical law is simple: to give is to live. And the root experience of love is, I think, the amazing realization that another person actually means as much or more to me than myself. Based on this realization, love as the gift of self is an act of faith because it neither expects nor requires anything in return. As the Buddhists say, true love implies that we "cease desiring."

This paradoxical discovery of a better self includes the sense of elevated purpose that allows us to develop our gifts and talents. One does not seek such discovery, for this would defeat the genuine call of love for others. But such love always brings with it a sense of meaning, noble purpose, well-being, and creative endeavor. I recall the symbol of an Episcopal secondary school in New Hampshire. On top of the school shield is the pelican, a medieval symbol of Christian love because this bird will pluck its chest vein if need be in order to feed its offspring. Below the pelican are crossed swords, indicating that love must be accompanied by courage and justice. Last comes the open book, *libris*, indicating that studying only has meaning if it is an outgrowth of loving purposes. We should study and learn not for selfish reasons, but because we love unselfishly and realize that effective love is intellectually demanding.

Each of us, from the springtime of youth to the winter of old age, is free to experiment with the way of love and discover its truth. Reach out in loving-kindness to the person by you on the left or the right, whether at school, at home, at work, in your community, or in the hospice as you lay dying. Reach out in loving-kindness not because this will enhance your sense of purpose and self-esteem, although it probably will, but simply to help others in the spirit of generosity. Reach out in loving-kindness not because you expect any "payoff" in reputational gain or reciprocal response, although unselfish connections and relations may well result from such actions and give rise to a surprisingly deepened community. Forget yourself by pretending that you are a little demented, love others for the sake of others, and let all the unintended aftereffects such as improved well-being and community take care of themselves.

The giving of self in thankful celebration of the lives of others, and in concern for their well-being and in attentiveness to their needs, describes in general terms all the forms of higher love. Such love differs from the lower forms of "love," by which we usually mean motivations that give rise to a concern for others only as means to the realization of selfish agendas. Such love may appear impressive but it is false and ephemeral. In all higher love, we forget our own agendas and discover others as independent centers of value. In stepping away from our old narrow selves to wider spheres of love, we will experience some sense of loss, but that will be entirely forgotten in the exciting discovery of our deeper nature. Because spiritual growth—that is, growth in love—can never be coerced, each individual must grow into love at his or her own pace.

Love can take so many forms. As a professor, I am always impressed at graduation ceremonies when family members and friends gather joyfully around a loved one who has a new degree in hand. Their loving delight in his or her successful completion of studies and in the start of a new stage of life is unmistakably written on their faces and heard in their tone of voice. Such joy is palpable, meaning that it is easily observed and felt by anyone awake to those around them. Here love takes the form of *celebration*, and we need these occasions to remind us that our life is a blessing. The extent to which well-wishers can forget their selfish concerns and enter a playful freedom from the usual anxieties of life is remarkable. In times of celebration, we give ourselves and we discover ourselves.

Love can take the form of active *compassion* when someone is suffering and needs support. Compassion includes responsive helping behavior. It is an emotional state with practical consequences. In times of compassion, we give ourselves and we discover ourselves.

Love can take the form of *forgiveness* when someone needs to be reconciled with the community, with loved ones, or with a nation after making mistakes of significance. Everyone who is truly apologetic deserves to be forgiven. Forgiveness is powerfully articulated in the Lord's Prayer (Matthew 6:10) and in Psalm 103:8–13, where we are told that "The Lord is merciful and gracious." The theme is brought forward in many New Testament parables, such as the parable of the unforgiving slave (Matthew 18:21–35), and in the epistles, such as Romans 14:5–12. In times of forgiveness, we give ourselves and we discover ourselves.

Love can take the form of *care* when someone falls ill and has needs that only others can meet. Every day, family caregivers tend to the needs of children, of older adults, and of loved ones generally. Professional caregivers, from the health professional to the social worker, are trained to give competent care as needed. We are all dependent beings at many points in the journey of life. In times of care, we give ourselves and we discover ourselves.

Love can take the form of *companionship* when solitude grows tiring. The simple experience of being with another in friendship is a form of love. Breaking bread together is companionship, and thus the sharing of meals has meaning. Offering wisdom in quiet confidential conversation is companionship. Attentive listening in which one tries to be fully present rather than distracted is companionship as well. In times of companionship, we give ourselves and we discover ourselves.

Love can take the form of *correction*. It always affirms the value of all others, but it will not affirm hatreds and harmful actions. Love leads through courage, strength, and wisdom in effectively restraining evil. It should not be arrogant, self-righteous, or in many cases, overly certain of its perspective on events. And yet love requires us to deter wrongdoing after motivational self-examination and conscientious discernment. Love is ready to skillfully confront behaviors that are self-destructive as well as destructive of others. As psychiatrist M. Scott Peck writes, love must be willing to take "the risk of confrontation."[1] But in confrontation, love must never give way to malice. Love often confronts only indirectly through questioning, example, suggestion, and the gentle forms of shaping influence; and yet maleficence in its fervor will require something more direct. In times of tough love, we give ourselves and we discover ourselves.

Giving ourselves in unselfish love is transformative. Religious traditions have always captured this insight in their narratives. The *Rg Veda*, for example, a foundational Hindu text, introduces the concept of *Rta*, or sacrifice, into cosmology and human growth. Sacrifice of the old self is a necessary prerequisite for any subsequent development. This constitutes a law of nature, just as the shell of an egg must be broken through in order for a bird to be born and fly, or a cocoon must be ruptured for the butterfly to emerge. Christianity speaks of *kenosis*, a Greek word meaning literally "emptying" in spiritual generosity to open the heart of another (Philippians 2:6–11). The self that is unloving is untransformed. It is like the snow and cold wind of a bitter winter. When the weight of selfishness is lifted, fragrances of spring make us into something more and better. This is the universal law of renewal. One finds a new self in deep forgetfulness of the old. Everything that is objectively good, both ethically and spiritually, is grounded in this *kenotic* renewal. Buddhism takes the view that the possessive ego and selfishness are the root of all suffering, and that one can overcome selfishness through meditative techniques and compassionate love.

No matter what we try to do in life, and no matter how successful and impressive our external talents are, in the absence of love nothing is worthwhile, and nothing will last. The kingdom of God becomes real to the extent that God's love and justice become our own.

This book is divided into three sections, the first of which takes up the question of what we mean by "unlimited love." The reader will note that at times, when I am thinking of this love in "big picture" terms as an ultimate reality underlying all that is, I use the uppercase: Unlimited Love. The second section focuses on social, scientific, and evolutionary perspectives on human altruistic motivations. A final section touches on the beginnings of the field of unlimited love research.

Human progress through the increased practice of love is the only alternative to the stagnation of egoism and conflict. Perhaps Unlimited Love is the Master Poet behind the universe, fostering love in a still incomplete and chaotic human world, and ready to change the hearts even of those who in their narrow loyalties have yet to discover the Beloved Community of all humankind. Unlimited Love may be a real energy that draws forth latent human possibilities. Are our human potentialities for love much greater than most of us think? Perhaps this is delusion and fancy, but if no such poet exists, there are still arguments to be made that the direction of human development has been toward greater cooperation and that love is an evolutionary necessity.[2]

Yet we still live in a world of violent ethno-nationalist conflicts, which continue to mar the prospects for the human future. Why are we so susceptible to hate-filled indoctrination and negative ideologies or beliefs that lead to intergroup hostility? This human tendency, with its deep evolutionary and historical roots, becomes particularly frightening in the context of the modern technology of mass destruction through biological, nuclear, and chemical agents.[3] But such circumstances seem to point toward a potential renewal of our sense of a common humanity as the only way to achieve a salutary future for our children.

John Templeton suggests that love is not a creation of people but rather people are a creation of love. With this perspective he reminds us to be open to the metaphysical reality of love underlying and sustaining the universe and human progress.

PART ONE
WHAT IS UNLIMITED LOVE?

1

UNLIMITED LOVE

AND ULTIMATE REALITY

U NLIMITED LOVE is a mysterious point of convergence between all worthwhile religions and is deemed the essential aspect of a presence in the universe that is infinitely higher than our own. While some view this unselfish love for all humanity as merely a human moral ideal, I view it in metaphysical terms as the ultimate reality that underlies all that is, and which can transform our limited and broken lives into journeys of remarkably generous service. If Unlimited Love really does describe ultimate reality, then we need not be so surprised by people who do so much to serve humanity while claiming that they are inspired by a loving energy in the universe. Unlimited Love is God's love for us all.

Human love sometimes shines brightly and is enhanced by the pure brilliance of Unlimited Love, but it is usually a dim reflection of love at its heights. And yet we encounter astounding examples of unselfish human love and sacrifice, suggesting either that our capacity for love is much greater than we might imagine, or that we can be lifted up into the ultimate reality of Unlimited Love. We marvel at those who achieve higher degrees of unlimited love and derive from their example a hope for something much grander in the human future.[1] By analogy, when we hear a composition by Bach or Mozart, we are astounded that any human being could be capable of such musical creativity. We wonder if they reached such heights on their own, or if God reached into their lives with divine creativity. Were they harbingers of a human future in which we will all be able to discover such astounding creative genius within?

What amazing creative and loving capacities rest untapped in the human being, or in other higher beings that may exist somewhere in the universe?

Is there a dynamic synergy possible between human and divine love? In the eighteenth century Charles Wesley, one of the founders of Methodism, wrote these hopeful words about God's love dwelling gracefully in the humble: "Love divine, all loves excelling, joy of heaven, to earth come down, fix in us thy humble dwelling. . . ." Did Wesley capture a perennial spiritual dynamic?

In every age, people ask if unlimited love is ultimately real, and if it can touch us. Many people devote their entire lives to a spiritual quest, as if they were "born to run" until they can find some answer to a dominant question of the human journey: Is this ineffable love, described by mystical poets and by many of those "called" to lives of compassionate service and kindness, a reality? Can Unlimited Love enable us to break through our evolutionary limits and thereby excel in love just as an Einstein did in physics or Shakespeare in writing? Our complex brains, unique imaginations, communicative abilities, reasoning powers, moral sense, and spiritual promptings can give rise to that which is extraordinary in human life as we resonate with, or even participate in, divine love.

On an existential level, exemplars of love pose a question: Is generative love for others the main purpose of our life, the only enduring source of meaning and dignity, and the basis for lasting self-esteem? Many psychologists, religious thinkers, and philosophers have answered this question in the affirmative. Moreover, in any honest and profound struggle for meaning in life, many everyday people understand this to be true.

We limited human creatures live somewhere "betwixt and between" egoism and unlimited love. Some of us live entirely for self, pursuing selfish agendas, but ultimately we find no final fulfillment in this. Some live astoundingly generous lives toward friends or kin, but are intolerant of outsiders and overindulge these loved ones. Many of us, however, are able to deeply affirm and serve all humanity while also nurturing the near and dear who are by human nature benevolently structured into our lives as part of that humanity. It is this last image of human fulfillment that is most legitimate ethically and spiritually. We should love all humanity, and within this love recognize that some people are placed by family and friendship in unique moral proximity.

Love for all humanity without exception is not innate. It was not a visible ideal among the Greeks, who could not see beyond the city-state (*polis*), and even within the city-state, friendship (*philia*) was king. While there was in antiquity a weak notion of *philanthropia* or "love for humanity," this did not enjoy any fuller development until the late Stoics and it did not apply to humanity as a whole. Judaism too was insular in its beginnings but of great historical significance in that it introduced the notions of hospitality to aliens (non-Jews) and of moral obligations to humanity as a whole. These ideals also remain significant in Islam. Buddhism and Christianity would introduce the remarkable ideal of loving even our enemies. Love for all humanity in this intensive form of love for enemies is especially challenging to any human being. Mahatma Gandhi, a Hindu, wrote, "it is no non-violence if we merely love those that love us. It is non-violence only when we love those that hate us. I know how difficult it is to follow this grand Law of Love. But are not all good and great things difficult to do?"[2]

We are not "hard-wired" to love all humanity, and this form of love has sometimes been reduced to a thin veneer covering a seething cauldron of human hatred and group conflict. With the power of weapons of mass destruction, we must now learn the lesson of love for all humanity or perhaps suffer enormously. Every age can be defined ethically and spiritually by how well it teaches and implements the ideal of love—and love implies justice—for every human being without exception.

Unlimited Love is an uncreated and perfect energy; human love is created and imperfect. The purpose of every human life is a movement toward greater love, no matter what the circumstances. Throughout the course of life we are all equally called to continual growth in love, even as we suffer setbacks along the path. The goal was eloquently defined long ago by the Hebrew prophet Micah: "What does the LORD require of you but to do justice, and to love kindness, and to walk humbly with your God?" (Micah 6:8).

Some readers may be disappointed to discover that unlimited love is not about the romantic phenomenon of "falling in love," in which every hormonal and aesthetic human propensity powerfully drives the self to an overwhelmingly positive assessment of another in which the beloved becomes perfection incarnate. This deeply evolved and "hard-wired" infatuation is reproductively necessary and, therefore, in the succinct words of one philosopher, it is a "trick of the species." Not all romantic infatuation lasts for only a few months or years, although it tends to be fleeting. Some people may "fall in love" for a lifetime, although one suspects that the reason such relationships last has to do with the underlying support of other forms of love, such as compassion and friendship based on common interests, like children. Regrettably, our culture too often reduces the meaning of love to romantic infatuation and nothing deeper. What we focus on here is not a "fall" into love of the perfect, which is altogether natural, easy, and delusional. Our focus is instead on a difficult, learned ascent that begins with insight into the need for tolerance of ubiquitous human imperfection, and matures into unselfish concern, gratitude, and compassion.

Some others will be concerned to find that we plan to explore unlimited love as the interface of religion, ethics, and science. Religion, they will rightly assert, is the cause of at least as much evil as good in the world. Religious arrogance underlies much conflict in our contemporary world, and there are even some extremists who would rather see those with differing beliefs dead than alive. Religion is an ambivalent force that can tap both the love and the hate that lie within us. Those for whom religion spurs love report perceptions of a source of Unlimited Love that can quicken the spirit of beneficence in human events: they seem to have a deep emotional attunement of gratitude for each and every life, and to affirm that every person without exception can be loved. They are ostensibly forgiving and patient, and possess great energy in the service of others. Are such people gifted by God? Or have their imaginations, desperate for a sense of purpose, triumphed over reason? Is it our human destiny to emerge from our human origins in "selfish genes" and intergroup conflict to a world of greater and greater love? Is Unlimited Love up ahead luring humanity into a future of loving-kindness for everyone, as well as for nonhuman species, both lower and higher? Is there some unfolding purpose to the evolution of our species that is related to love? In the words of Paramahansa Yogananda, best known for his *Autobiography of a Yogi*, "As a mortal being you are limited, but as a child of God you are unlimited." He asks if we might eventually transcend our "mortal consciousness of limitation" and achieve unlimited love, for such "divine love is without condition, without boundary, without change."[3]

Even the most abiding human love for others cannot ignore the strategic necessity of imposing limits on recalcitrant hateful behavior. Reinhold Niebuhr asserted that those "children of light" who wish to bring egoism under the control of love must have the wisdom and cunning of those who would assert selfishness, but none of their malice.[4] Given the reality of a human nature open to narcissism and hatred, love does not disparage strategic limits that are required in response to malice. We deem it acceptable to limit violent and abusive behavior, and to defend the innocent

against assault. This is why fortitude, courage, and endurance are important aspects of the expression of love, under some circumstances. When-
ever possible, conflict must be transformed into reconciliation. Nonviolent
means are always to be exhausted before alternatives are considered.[5]
Whether resistance to evil is nonviolent or not, our actions can be motivated by a love that remains open to forgiveness and reconciliation.[6]

The realities of excessive ambition and egoism highlight the extent to
which "tough love" is an absolutely essential aspect of unlimited love. We
should not naïvely forget the strategic necessity of love in its tough form in
response to sinful self-assertion.[7] Human evil is as real as human love, and
love is called upon to contain the fury of rudeness, greed, and hatred until
a time comes, if ever it does, when human nature becomes something
much better than it is. Remarkably, the young Lutheran pastor Dietrich
Bonhoeffer decided that the only loving thing to do in Nazi Germany was
to participate in the "officer's plot" to assassinate Hitler. Bonhoeffer was
executed by the Nazis in 1945 for his part in that failed plot. He appears
from his writings to have maintained his grounding in *agape* love throughout these years of difficulty.[8] Conclusions like Bonhoeffer's should be
reached only under extraordinary circumstances, and with great reluctance.

Is Unlimited Love Real?

The discontinuities between pure Unlimited Love and human love are obvious, but there are hints of continuity as well. Parents can feel utterly
absorbed by love for their children, seeming to naturally feel that the lives of
their young are as important as their own, and probably much more so.
Hence, parents routinely sacrifice their interests for the prosperity of
offspring. This love does appear to be innate to human nature, however
much it can afford to be tutored in wisdom and efficacy. The significance
and purity of such love is not compromised by a substratum of genetic
continuity—that is, "selfish genes." Indeed, all human compassion and generous love for others have their earthly roots in this powerful evolutionary
axis of parental love, or so I will later argue. Does the unlikely presence of
such compassionate and caring love on earth point to divine origins?

While we are acquainted with hatred and violence, we are equally
acquainted with neighborly acts of compassion and care within our families,
and between friends and colleagues. We see humanitarian groups all over

the world rushing in to help those who have been in the way of natural disaster or human harm. The gift of personal time and sacrifice is often heroic, and the loving-kindness of strangers is a marvel to behold. Even in the worst times, such as the Columbine or the World Trade Center tragedies, we are impressed by the helpers and we are reassured the entire world is, on some level that has not yet completely broken through the surface, a Beloved Community. We see those who, in addition to organizing voluntary associations of helpers in response to every conceivable need, also recognize that love must do justice. (No matter how much we live by the law of love, there are often limits to what we can accomplish singlehandedly or in associations, and therefore we must strive to correct social and distributive injustices.)[10]

The essence of all true spirituality, religion, and virtue is continual growth in the direction of Unlimited Love.[11] Such enduring affection can be distinguished from fleeting emotional experiences, no matter how intense these might be, or how much they result in limited periods of enthusiasm. However one wishes to define the human "soul," it serves metaphorically to focus attention on whatever is ultimate and essential in being human. The "soul" is a complex place where our loving emotions struggle for ascendancy. Fear, terror, anxiety, hatred, bitterness, violence, envy, resentment, greed, and pessimism try to roam unchecked, strangling the life out of our lives. Spirituality refers to the ascendancy and dominance of another set of emotions associated with love. The fruits of such love include peace, joy, gratitude, kindness, care, forgiveness, and concern. All true spirituality is a form of love, and is a matter less of creed than of affection.[12] All true positive spiritual transformation involves a shifting of the emotional balance toward Unlimited Love. In the words of William James, all genuine religious experience involves "an assurance of safety and a temper of peace, and, in relation to others, a preponderance of loving affections."[13] The ultimate expression of love is love for all humanity, and for all that is. Any spirituality or religiosity that betrays the ideal of growth toward Unlimited Love is simply negative and dysfunctional. A statement such as this requires some closer attention to definition.

"Unlimited Love" Defined: Initial Thoughts

By now you may be looking for a clearer definition of love than any I have yet provided. "Love" is a confusing term, so much so that it is often replaced by the word "care." If common use of language is revealing, some seem to assume that love must involve sexual expression. This is incorrect because friendship is a form of love, as is volunteerism or the care that parents give to children. We erroneously use the word "love" to describe a caprice of a few days' duration or a sentiment devoid of esteem. The Greeks were more careful to make linguistic distinctions. They had a myriad of words for love: *eunoia* refers to good will or benevolence, *physike* to kindness toward people of one's own race, *xenike* to kindness toward guests and strangers, *erotike* to sexual desire, *eros* to impassioned attraction, *philia* to friendship, *storge* to tenderness, and *agape* to a disinterested affection.

Agape would be taken up by emergent Christianity and identified as the essential nature of God, a divine limitless love. This affectionate love for all humanity seems to have at least some place in all major religious traditions of the world. Unlimited love is a way to speak of *agape* love in a contemporary world where most people cannot be expected to be versed in Greek or in the history of ideas, but sense that there is an abiding energy of love in the universe, and that this love affirms them despite their failures and imperfections. The expression "unlimited love" seems to capture the essence of *agape*, free of a narrow association with any one faith tradition, and should appeal more broadly across cultures, languages, and academic disciplines or fields. Sir John Templeton coined the term "*Unlimited Love*" in an essay entitled *Pure Unlimited Love: An Eternal Creative Force and Blessing Taught by All Religions.*

Agape love has been described as "limitless" or "without limits" in the theological literature.[14] Templeton is correct in suggesting that "unlimited" is the best word to describe *agape* love. It suggests a form of love that rises above every conceivable limit to embrace all of humanity in joy, creativity, compassion, care, and generativity; it lies at the heart of all valid and worthwhile spiritual, religious, and derivative philosophical traditions; it is often associated with a divine presence that underlies the cosmos and makes life a meaningful gift. One purpose of religion is to provide a larger theological canopy for such love, to translate this ideal into some sacred narrative, and to encourage its uncoerced and gratuitous expression. Anyone who senses the limits of living in anger, hatred, egoism, unkindness, sensate hedonism, rudeness, greed, grudge-bearing, insular tribalism, and violence at least perceives some hint of the ideal of unlimited love.

Unlimited love does not require a response. It can be entirely unwarranted

and undeserved, as well as dismissive of the purported imperative of reciprocation. While it "seeketh not its own," it accepts and delights in gratitude
and other responses, however much it does not require or expect them.
Unlimited love involves a daring, free, compassionate, and urgent dream
that explodes through all the requirements of benefit to self.

It is easy to think of love as an energy, the sum total of which defines
the goodness of any society. The total love energy of a society is, for the
most part, a measure of the everyday compassion and helping behaviors of
ordinary people who are good neighbors and who have an abiding concern
for those around them as well as for the very neediest. These people are
quite plentiful, especially in times of catastrophic events in which our common vulnerabilities are magnified. Everyday altruists hold "perceptions of
a common humanity."[15] As Pitirim A. Sorokin pointed out in his classic
1950 study of good neighbors and saints, the great apostles of the most
creative levels of altruism are relatively few and cannot provide anything
like the sum total of love energy that a society requires to thrive.[16]

What, then, of the word "unlimited"? The word "limit" derives from
the Latin *limis*, which means "boundary." Oxford's dictionary defines a
limit as "a point, line, or level beyond which something does not or may
not extend or pass." "Unlimited" means without limit, or with no restric-
tion whatsoever. Unlimited love is love for all humanity and, on a lesser
ontological level, for all living creatures. "Unlimited love" means that there
will be no insulating boundaries drawn to separate "them" from "us," that
love for the neediest stranger must deeply and honestly challenge any
overindulgence of the near and dear, that even the most vile enemy must
be forgiven at some level.

John Templeton notes that "unlimited love was called *agape* by the ancient
Greeks to distinguish the divine love from earthly emotions. Unlimited
love means total constant love for every person with no exception."[17] Such
love, argues Templeton, is productive of health and peace in the world.

Unlimited love echoes the Jewish notion of *hesed* ("steadfast love") and
the Buddhist ideal of *karuna* ("compassion"). It had rough equivalents in
Islam, Hinduism, Taoism, Confucianism, and Native American spirituality.
Such love was the centerpiece of works by Carl Rogers, Rollo May, and
Erich Fromm.

The question, of course, is how to harness the creative energy of unlimited love. We need to better understand the obstacles to unlimited love,

and how to overcome them. Delving into the history of ideas and practice of love may be helpful, but the frank reality is that unlimited love has not been fully realized even in those traditions that most eloquently extol it. Thus, there is a need for immense new knowledge of love that is based on the best scientific methods and that can, therefore, move our understanding forward, allowing for more effective pedagogy of the young and the old, and the transformation of culture and society.

What, then, do we mean by unlimited love? *The essence of love is to affectively affirm and to gratefully delight in the well-being of others; the essence of unlimited love is to extend this form of love to all others in an enduring, intense, effective, and pure manner.* In addition to being understood as the highest form of virtue, unlimited love is often deemed a Creative Presence underlying and integral to all reality. Participation in unlimited love is considered the fullest experience of spirituality, giving rise to inner peace and kindness, as well as to active works of love toward all humanity. Depending on the circumstances of others, love is to degrees expressed in a number of ways, including empathy and understanding, generosity and kindness, compassion and care, altruism and self-sacrifice, celebration and joy, and forgiveness and justice. In all these manifestations, unlimited love acknowledges for all others the absolutely full significance that, because of egoism or hatred, we otherwise acknowledge only for ourselves.

One definition found in the theological literature that I think captures the root of the experience of love is this:

> By love we mean at least these attitudes and actions: rejoicing in the presence of the beloved, gratitude, reverence and loyalty toward him. Love is rejoicing over the existence of the beloved one; it is the desire that he be rather than not be; it is longing for his presence when he is absent; it is happiness in the thought of him; it is profound satisfaction over everything that makes him great and glorious.[18]

Rejoicing, gratitude, reverence, and loyalty are all constituent elements of unlimited or *agape* love. Love is gratitude or thankfulness for the existence of the other; it is a reverence that seeks not to absorb the other or refashion him or her as an image of the self; it is a loyalty that would rather allow the self to be destroyed than have the other cease to exist.

The obvious opposites of unlimited love are hatred and destructiveness. The more subtle opposites are intolerance, snobbery, and a general solipsism that is blind to human equality. Thus the moral theologian Gene Outka defines *agape* in terms of "equal regard."[19] Love means overlooking all of the perceived aspects of another that one finds simply intolerable and cause for an attitude of disdain. It means seeing through to the immense worth of each and every person as a living human being and affirming them equally as such with one's full heart.

If we look to the New Testament, we find that *agape* or unlimited love is first and foremost God's love, or God who is identified as love (1 John 4:16). Unlimited love is deemed the primordial energy in God's eternity, and the prime force underlying both creation and redemption. This love is a gift that desires to be mirrored in us though the synergy of our naturally evolved propensity for compassion and the effects of divine grace working directly on the individual. To be created in God's image means that we are created *for* love *by* love. We know that we are loved by Unlimited Love, and therefore we are able to love. And yet questions remain: How much of divine love is reflected within human affect and nature? How can we better understand the experience of Unlimited Love as a divine energy that makes the substrate of human nature into something even better?

THREE CLARIFYING EXAMPLES

This chapter has been rather abstract thus far, and it is time to consider some concrete examples of proximate unlimited love in human lives. Three stories are introduced here to clarify just what love it is that we are talking about. The first two are contemporary.

Most of the unselfish love energy in society is found in all of the small opportunities that average folks have to be good neighbors: visiting the sick, helping a friend, or offering to step in for a caregiver. These small things, done in the spirit of love, mount up and shape the world. One of my favorite examples of unlimited love expressed in ordinary lives is the inspiring story of Patty Anglin and her family.[20]

The Anglins have fifteen children, seven of whom are biological and eight of whom are adopted. All of the latter have "special needs": serious medical problems, ranging from cognitive deficits to the absence of limbs. While there have been many challenges, the couple writes that these have been overcome by their experience of answered prayers and miracles of coincidence. They describe their family as "a sort of mini-United Nations," with children from a myriad of ethnic and racial backgrounds. The daughter of a physician and a mother who was "the most unselfish person I have ever known," Patty Anglin grew up in the missionary fields of Africa. She and her husband now live on a farm in Wisconsin that is named "Acres of Hope," spreading what they perceive as God's love, one child at a time.

A second story is that of a poor and abused Irish girl from the Dublin slums by the name of Christina Noble. Her book, *Bridge Across My Sorrows*, begins with a vision:

> I came to Vietnam because of a dream I had almost twenty years ago. The dream told me to work with the street children of this poor, jangled, disease-ridden country. You might laugh at that. You might say it was nothing but a dream and that only someone who was Irish would act on a dream as if it were a message from God. And you could be right. After all, my coming here was not anything I could explain then or anything I can explain today. I had a dream—a vision, if you will—that ordered me to Vietnam. That is all.[21]

Christina's dream was to work with street children, for in it she had seen "a little girl reaching towards me for help." She arrived in Vietnam in 1989, almost twenty years after her dream, "a middle-age woman with no education, no money, and no real idea of what I was going to do in Vietnam." She left behind a life she considered unaccomplished, except for the birth of her three children. Now she would discover if the dream was only a dream, or a vision to direct her life. In the dream, Christina knew that the girl reaching out was hungry, frightened, and without family:

> Behind her were many other children, all rushing towards me and all crying for me to help. But I couldn't reach her. Then she was gone and all that was left was grey smoke that twirled and twirled as the wind moaned. Through the smoke appeared a light, a great white light, and within seconds the light evolved into letters. Then the word "Vietnam" was burning across the sky in brilliant, almost blinding lights.[22]

Now in a filthy hotel in Ho Chi Minh City, Christina saw two little girls dressed in rags playing in the dirt across the street. Only they were actually "grubbing for ants" and eating them. The girls called out to her, "Give me. Give me." Christina pulled back, for the sight evoked for her memories of her own childhood in the Irish gutters.

But one of the little girls reached out to her, wanting the "touch of another human being." Remarkably, "Her hands and her expression were those of the girl in my dream." Christina reached out and embraced her, and her life would forever be changed with that momentous decision, for she would work with these children who live as she had lived in Dublin: "This poor and crippled country would be the place of my salvation, the place where I would regain hope and rebuild my life. Here I would stay. Here I would find happiness. I knew that I would never leave."[23]

In Christina we see an example of a person who overcame horrible personal circumstances through love. Born in Dublin in 1944, her father was a violent alcoholic, and her mother, whom she loved, died when Christina was a little girl. She grew up in a cruel orphanage and escaped to become destitute on the streets of the city. At sixteen she was gang raped by four men. She learned to work hard at tough jobs, started a small catering business, and was married to a violent and abusive husband with whom she had three children. Overworked and in despair, she had her dream of the children of Vietnam, and she found all hope in a determination to someday work with them. Her work as "Mama Tina" in Ho Chi Minh City is remarkable for its success. She felt throughout that she was doing exactly what God wanted her to do, and this gave her confidence. "Mama Tina" succeeded in creating a major center for the care of street children that is both a hospital and a social center. Children, most of them malnourished, are treated medically, and then moved to a residential care facility. At any given time an estimated seventy-five children are under inpatient residential care, and a thousand are treated each month on an outpatient basis.

One of my favorite historical stories of love is that of the eighteenth-century American Quaker John Woolman, who wins my prize as one of the most courageous and effective "tough" lovers in world history. An undistinguished colonist, he traveled to Quaker meetings across the colonies, witnessing to people one by one about the evils of slavery. "My heart was tender and often contrite," he wrote, "and universal love to my fellow creatures increased in me."[24] The modern antislavery movement began precisely at the moment when, as a matter of conscience, Woolman could no longer assist his employer in the sale of a slave. After convincing his fellow colonial Quakers to give up slavery, Woolman went on to spread his message in England, where he died after a short period of intense endeavor. Many Quakers both in the colonies and in England had been slave-owners. Woolman set about the task of creating change through the art of tough love. He was confronting evil in the spirit of love, defying a convention that he found intolerable, and succeeding in quiet one-on-one conversations. He visited his fellow Quakers individually, farm after farm, for most of the two decades of his adult life. As Robert E. Quinn, a leading scholar on contemporary leadership skills, writes in his study of visionaries and "deep change," Woolman did not criticize people or anger them, but "by 1770, a century before the Civil War, not one Quaker owned a slave."[25] Due to the efforts of one man who seems to have approached the ideal of unlimited love as nearly as anyone, the Quakers were the first religious group to denounce slavery. If there had been a John Woolman in every religious denomination, perhaps the institution of slavery could have been abolished without the need for civil war. A single visionary individual committed to change under the power of unlimited love can make a difference in the world.

These three stories are about ordinary people with no special training, education, or social background who achieved remarkable outcomes simply by determining to live by faith in Unlimited Love. None of them would wish to claim that he/she understood or manifested the mystery of Unlimited Love, although it does seem that each discovered some resonance or synergy with this higher love that elevated them to noble purposes. Whether through a sense that every child is equally God's own, or through a vivid dream that was firmly ensconced in memory, or through a moral dictate that slavery is beneath the God-given dignity of a human soul, each achieved miraculous things.

Do these stories help persuade those who reject the possibility of genuine unselfish love for anyone at all, let alone for all humanity, of the existence of Unlimited Love? The skeptic asserts that those who serve humanity are really pursuing reputational gain within the constraints of universal egoism and duplicity.[26] Or perhaps such good neighbors are motivated by the hedonic satisfaction felt after a task well done. The world needs skeptics, although I will present enough scientific evidence in the next section of this book to suggest that such skepticism is not empirically supportable. We need a new saying, "Scratch an egoist and watch an altruist bleed." But we turn first to the pioneering work of Pitirim Sorokin, who is significant for suggesting a five-dimensional measure of the quality of love. Sorokin's science of love will provide more clarification of the meaning of love as it is considered in this book, as well as indicate how empirical methods might apply.

2

THE MEASURE OF UNLIMITED LOVE

IN THE WORK OF SOROKIN

I F WE LEAVE the domain of love only to the philosophers and theologians, the tough-minded empiricist will dismiss all of its manifestations as "soft." Science is, with regard to cultural impact, the dominant shaping force, and it is novel research that promises to resurrect our con*fi*dence in the power of love.

Most scientific studies on other-regarding love consider it under the broader term of "altruism," which emerged in modernity as a secular replacement for *agape* or charity. Altruism, which can be grounded in love, refers to genuinely motivated helping (beneficent) actions. Such altruism might be a manifestation of a purely rational Kantian imperative ("This is what pure reason and duty dictate"); a sense of self-identity and expectation associated with social role or super-ego ("This is what someone like me is expected to do"); or of an innate compassion that reacts to the needs of others. The latter alone implies the affective quality of love.

Altruistic love was best studied by Pitirim A. Sorokin (1889–1968), a towering figure in twentieth-century sociology.[1] As a young man in Russia, he was imprisoned first by the Czarists and then by the Bolsheviks; he concluded that Czarist prison was the more comfortable of the two. After immigrating to the United States in 1923 to teach at the University of Minnesota, he went on to become the founding chairman of the Department of Sociology at Harvard University in 1931. In 1945, anxious over the human condition in the wake of World War II and Hiroshima, he founded a program called the Harvard Research Center for Creative Altruism. In his autobiography, *A Long Journey,* Sorokin expressed pessimism about the potential for political efforts to bring about peace without the "notable altruization of persons, groups, institutions, and culture."[2] He was hardly sanguine about the role of extrinsic religion, because his own studies indicated that a "purely ideological belief in God or in the credo of any of the great religions" rarely results in more altruistic behavior. He became increasingly interested in investigating "scientifically this unknown or little known energy" of love: "The phenomena of altruistic love were thought to belong to religion and ethics rather than to science. They were considered good topics for preaching but not for research and teaching."[3] In a voice that has since been heard by the positive psychology movement of the 1990s, Sorokin noted the tendency of scientists to focus research on the disease model:

> While many a modern sociologist and psychologist viewed the phenomena of hatred, crime, war, and mental disorders as legitimate objects for scientific study, they quite illogically stigmatized as theological preaching or non-scientific speculation any investigation of the phenomena of love, friendship, heroic deeds, and creative genius. This patently unscientific position of many of my colleagues is merely a manifestation of the prevalent concentration on the negative, pathological, and subhuman phenomena typical of the disintegrating phase of our sensate culture.[4]

The above point has not been lost to positive psychologists, although the contemporary movement does not yet recognize its resonance with Sorokin's legacy.

Sorokin's publicly disseminated statement describing the Center for Creative Altruism makes explicit his assumption that "none of the prevalent prescriptions against international and civil wars and other forms of interhuman bloody strife can eliminate or notably decrease these conflicts." He continued:

> Our second assumption was that this unselfish, creative love, about which we still know very little, potentially represents a tremendous power—the veritable *mysterium tremendum et fascinosum*—provided we know how to produce it in abundance, how to accumulate it, and how to use it; in other words, if we know how to transform individuals and groups into more altruistic and creative beings who would feel, think, and behave as real members of a mankind united into one intensely solidary family.[5]

As one would expect, Sorokin's first task with the Research Center was to collect all the useful data he could. In 1950, he published an important book entitled *Altruistic Love: A Study of American Good Neighbors and Christian Saints*. At the beginning of this impressive empirical work, Sorokin wrote that while "for decades Western social science has been cultivating, *urbi et orbi*, an ever-increasing study of crime and criminals; of insanity and the insane; of sex perversion and perverts; of hypocrisy and hypocrites," it has "paid scant attention to positive types of human beings, their positive achievements, their heroic actions, and their positive relationships."[6] This study looks at the typical characteristics of altruistic persons (sex, age, occupation, economic status, social position, education, and other features) and discusses their types, as well as how they became altruistic. Four years later, *The Ways and Power of Love* appeared. Arguably Sorokin's greatest work, it is a classic text because it transcends the limits of any particular era.

An Abundant Intellectual Passion

The Ways and Power of Love was published in 1954, when Sorokin was leading the Harvard Research Center in Creative Altruism. Like the author's earlier works on the dynamics of cultural change, this book received considerable attention. The philosopher Robert G. Hazo, for example, lauds *The Ways and Power of Love* in his classic 1967 work on the history of the idea of love, published in Mortimer J. Adler's series on Concepts of Western Thought:

> Sorokin treats love as a separate subject in a treatise devoted exclusively to it. His elaborate discussion and analysis of love, its causes and effects, its human and universal significance, its higher and lower forms, and its implications for other subjects constitute one of the most extensive treatments to be found in the systematic literature about love. *The Ways and Power of Love* is an ambitious attempt to subject analytical schemes to a phenomenon that Sorokin claims has both a human and a cosmic dimension.[7]

Indeed, nothing that Sorokin wrote over his brilliant career lacked this ambitious character.

While the sheer scope and dimension of love in the thought and science of Sorokin is impressive to the contemporary integrative mind, the increasingly rigid disciplinary style of most of his contemporary social scientists prompted a very mixed reception for the book. Sorokin challenged the narrow and technocratic aspects of sociology, which he felt was captive to small fragments of data while lacking in any larger systematic, cultural-historical framework that would make these data meaningful or interesting. Indeed, some sociologists viewed Sorokin to be at times more a philosopher of history than a methodologist—a critique that he would take as a compliment. Sorokin is probably best described as a creative and idealistic social thinker, devoted to scientific observation but with too wide-ranging an intellect to be content with a purely technical rationality. Sorokin was perhaps at his controversial best in laying out the cosmic as well as the human aspects of love. To fully appreciate his nonconforming genius, however, we must delve briefly into his distinctly Russian intellectual roots.

"Integral Knowledge" and Its Detractors

Sorokin inherited the Russian tradition of nineteenth-century thought associated with Nikolai Fedorov, Sergei Bulgakov, Feodor Dostoyevsky, Prince Peter Kropotkin, and Vladimir Solovyov, among many others. This distinctive intellectual tradition centered on the pursuit of *integral knowledge*, bringing together religious, psychological, ontological, cosmological, ethical, metaphysical, sociological, and biological knowledge. In addition to integral knowledge, these Russian thinkers ascribed to the key principle of *sobornost*, or "all-togetherness." They recognized the natural capacity for mutual assistance and cooperation, a theme articulated in the context of evolutionary science by Prince Kropotkin that today enjoys a significant renaissance in the related fields of game theory and evolutionary biology.[8]

In *The Ways and Power of Love*, Sorokin indicates the special influence of Vladimir Solovyov (1853–1900), a close friend of Dostoyevsky who synthesized philosophy and mysticism in his classic work, *The Meaning of Love*. Of love and its contrasts, Solovyov wrote:

> The basic falsehood and evil of egoism lie not in this absolute self-consciousness and self-evaluation of the subject, but in the fact that, ascribing to himself in all justice an absolute significance, he unjustly refuses to others this same significance. Recognizing him-

self as a center of life (which as a matter of fact he is), he relegates others to the circumference of his own being and leaves them only an external and relative value.[9]

Positively stated, Solovyov described the nature and value of love thus:

> The meaning and worth of love, as a feeling, is that it really forces us, with all our being, to acknowledge for another the same absolute central significance which, because of the power of our egoism, we are conscious of only in our own selves. Love is important not as one of our feelings, but as the transfer of all our interest in life from ourselves to another, as the shifting of the very center of our personal lives.[10]

Solovyov, like Sorokin, understood human love as a partial reflection of, and, at its height, a participation in, divine love.

It is against the background of Russian tradition, then, that Sorokin's broad concept of love can be most fully grasped. One can only imagine how foreign this tradition appeared to the academic world of western positivist sociology in the 1940s and 1950s, as it sought legitimacy through a narrower methodological model. Sorokin often wrote of the futility of sociology when stripped of deep integral reflection, for then it sacrifices interpretive insight and makes an idol of quantitative data. Although he was committed to scientific methods, he had a broader vision of his discipline than many of his contemporaries.

In *The Ways and Power of Love*, Sorokin considered love in seven aspects, following the tenets of integral knowledge.[11] The *religious aspect* of love identifies it with a Higher Presence, however symbolized in the great spiritual and religious traditions; the *ethical aspect* of love identifies love with goodness itself; the *ontological aspect* of love defines it as a "unifying, integrating, harmonizing, creative energy or power" that works in the physical, organic, and psychosocial worlds; the *physical aspect* of love is shown in "all the physical forces that unite, integrate, and maintain the whole inorganic cosmos in endless unities, beginning with the smallest unity of the atom and ending with the whole physical universe as one unified, orderly cosmos"; the *biological aspect* of love is evident in procreation and parental care. The sixth aspect of love is the *psychological*: "In any genuine psychological experience of love, the ego or I of the loving individual tends to merge with and identify itself with the loved Thee. The greater the love, the greater the identification." He views love as a "life-giving force" because of studies that show that people who are altruists live longer than egoists do. Love is also defined as "the loftiest form of freedom," for where there is love there is no coercion. Sorokin refers to the writings of St. Paul on this point. He was also conversant with a Russian contemporary, the theologian Nicholas Berdyaev, who emphasized that love nailed upon a cross compels no one. On the psychological level, Sorokin notes that love overcomes fear, as exemplified by the life of Gandhi, whom he much admired as a modern saint: "Love does not fear anything or anybody. It cuts off the very roots of fear." In a manner that brings to mind the various spiritual-ethical writings of the contemporary Dalai Lama, Sorokin associates love with "the highest peace of mind and happiness."

Seventh is the *social aspect* of love: "On the social plane love is a meaningful interaction—or relationship—between two or more persons where the aspirations and aims of one person are shared and helped in their realization by other persons." Sorokin quickly qualifies "aspirations" with the adjective "wise."

In *The Ways and Power of Love,* Sorokin focused mainly on the psychological and social aspects of love, but always with an eye toward its spiritual-religious aspects. In a passage that sums up his broad perspective on love and also captures the depth of his integral thinking, he wrote,

> Concentrating on these planes, however, we shall always keep in mind the manifoldness of love as a whole, because without its religious, ethical, and ontological aspects we cannot truly understand a "visible" part of this cosmos, its psychosocial empirical aspects.[12]

Here Sorokin is being true to his Russian intellectual tradition, with its wide-spectrum integral knowledge. Methodologically committed to new scientific knowledge that can move our understanding of love forward, he was also attentive to a wider cosmic context and to the fullness of human experience and history. Obviously, any scientific thinker working with these various contextual axes would have struggled for credibility in an era such as the 1950s, when the strictest diminution of the significance of metaphysical speculation was so dominant in major universities.

Sorokin's commitment to integral knowledge was addressed at an American Sociological Forum in 1959, which was convened by leaders of American social science and included such luminaries as Oxford historian Arnold Toynbee and the eminent sociologist Robert K. Merton, Sorokin's former student at Harvard. The fact that Sorokin received such attention is testimony to his continuing status, despite degrees of professional marginalization due in part to the rise of positivism. Joseph P. Ford devoted a presentation at this forum to the topic of "integralism" in Sorokin's work, pointing out his creatively integrative mind while acknowledging that the strict disciplinary emphasis of the time meant that considerable numbers of academics could not take Sorokin with appropriate seriousness.[13] In this same volume, Toynbee indicated the value of Sorokin's thinking on the history of civilizations that allowed him to see rhythms between alternating ways of life based on ideational, idealistic, and sensate values. Toynbee offered a critique that was largely appreciative, and suggested that Sorokin's sweeping integrative perspective was not unlike his own.[14] There was a resonance between Sorokin and Toynbee, and both went out of vogue as historians came to specialize on short periods of history devoid of rhythmic sweeping perspectives on the larger picture of the rise and fall of civilizations. Nonetheless, for those who approach history and thought with "big picture" questions in mind, Sorokin still makes interesting reading.

Integral knowledge, now a more acceptable notion in academic institutions, ran against the grain of academia for most of the second half of the twentieth century. Sorokin did pay a price. In one of the most famous infights in modern sociology, Talcott Parsons was able to absorb Sorokin's Department of Sociology into Harvard's then newly established Department of Social Relations in 1946, effectively superseding the more integrative Sorokin. In particular, Sorokin was criticized by Parsons and others for his ideas about cultural patterns and trends. Yet noted scholars of the era still referred to Sorokin as "the pre-eminent social philosopher of our age."[15] While sociologists continue to debate Sorokin's contributions, he was among the most prolific, creative, and widely translated sociologists of his generation long before he turned specifically to the topic of altruistic love.[16]

THE FIVE-DIMENSIONAL UNIVERSE OF PSYCHOSOCIAL LOVE

Sorokin's scientific mind led him to develop a remarkable five-dimensional model of love as a heuristic device for the articulation of core researchable questions. These questions still have importance to any research on love, and therefore should be highlighted here as they are developed in *The Ways and Power and Love.*[17]

Sorokin's first dimension of love is *intensity.* Low intensity love makes possible minor actions, such as giving a few pennies to the destitute or relinquishing a bus seat for another's comfort; at high intensity, much that is of value to the agent (time, energy, resources) is freely given. While Sorokin does not fully develop the different potential forms of intensity, his point remains clear. While the range of intensity is not scalar—that is, research cannot indicate "how many times greater a given intensity is than another," it is often possible to see "which intensity is really high and which low, and sometimes even to measure it."

Sorokin's second dimension of love is *extensivity:*

> The extensivity of love ranges from the zero point of love of one-self only, up to the love of all mankind, all living creatures, and the whole universe. Between the minimal and maximal degrees lies a vast scale of extensivities: love of one's own family, or a few friends, or love of the groups one belongs to—one's own clan, tribe, nationality, nation, religious, occupational, political, and other groups and associations.

Sorokin's extensivity resonates with the classic western discussion of the "order of love." How does one balance love for family and friends (the nearest and dearest) with love for the very neediest of all humanity?

Although the monotheistic faiths have appreciated the importance of special relationships such as family and friendship, they have asserted the centrality of love for humanity as a whole. Yet it is also possible for some-one to be so focused on the needs of all humanity that the importance of special relationships is missed. Various religious and philosophical tradi-tions have sought to resolve this tension in different ways. In this regard, Sorokin was especially concerned with in-group love because, as he argues toward the end of his book, insular group loyalty is rather typically the source of hostility against the out-group. With the advent of weapons of mass destruction, he feared such insularity might doom humanity. His pur-pose in doing research was in large part to better understand how insularity might be overcome. As an example of the widest extensivity he offers St. Francis, who seemed to have a love of "the whole universe (and of God)."

Sorokin next added the dimension of *duration*, which "may range from the shortest possible moment to years or throughout the whole life of an individual or of a group." For example, the soldier who saves a comrade in a moment of heroism may then revert to selfishness, in contrast to the mother who cares for a sick child over many years. Romantic love, he indicates, is generally of short duration as well.

The fourth dimension of love is *purity*. Here Sorokin wrote that our love is characterized as affection for another that is free of egoistic motivation. By contrast, pleasure, advantage, or profit underlie inferior forms of love, and will be of short duration. Pure love—that is, love that is truly disinterested and asks for no return—represents the highest form of emotion.

Finally, Sorokin identified a fifth dimension, the *adequacy* of love. Inadequate love is subjectively genuine but has adverse objective consequences. It is possible to pamper and spoil a child with love, or to love without practical wisdom. Adequate love achieves ennobling purposes, and is, therefore, anything but blind or unwise. Certainly love is concerned with the building of character and virtue, and will shun overindulgence. Successful love is effective.

These five dimensions of love allow us to ask empirical questions about how strength or weakness in one dimension varies with other dimensions. How intense, extensive, enduring, unselfish, and wise is any particular manifestation of love? Sorokin argues that the greatest lives of love and altruism approximate or achieve "the highest possible place, denoted by 100 in all five dimensions," while persons "neither loving nor hating would occupy a position near zero." Gandhi's love, for example, was intensive, extensive, enduring, pure, and adequate (effective).

In addition to allowing us to grade the total, five-dimensional quality of love in any individual, this method allows us to develop various ideal types, such as high intensity and very low extensivity, high intensity and very short duration, and low purity and short duration. Such an analysis invites researchers to classify "the types of love activities and of loving persons, and to learn which types and combinations are more frequent in a given human universe." The implementation of such an empirical survey would still provide useful data today. We might be surprised at what the findings would tell us about human nature; these findings might even run counter to the prevailing myth of self-interested individualism that seems to permeate the social sciences. We could also learn whether high purity correlates with long duration, and so forth. As Sorokin noted, "At present our knowledge of these relationships is rather meager." Nearly a half century later, this is still the case.

Sorokin did develop some interesting hypotheses regarding these dimensional relationships, all of which are worth pondering. For example:

The greater the five-dimensional magnitude of love, the less frequent it is in the empirical sociocultural world.

Other conditions being equal, the intensity of love tends to decrease with an increase of its extensivity or the size of the universe of love.

The intensity of love tends to decrease with an increase of duration, when the love expenditure of a given person is not correspondingly replenished by an inflow from other persons or other sources, empirical or transcendental.

Intensity, purity, and adequacy of love are somewhat more frequently associated positively than negatively or not at all.

Adequate love is likely to last longer than inadequate love.

THE THEOLOGICAL FEATURES

Of special interest to Sorokin was the love of figures such as Jesus, Al Hallaj, Damien the Leper, and Gandhi. Despite being persecuted and hated, and therefore without any apparent social source of love energy, they were nevertheless able to maintain a love at high levels in all five dimensions. Such love seems to transcend ordinary human limits. Sorokin argues that this seems to suggest that some human beings do, through spiritual and religious practices, participate in a love energy that defines God.

Sorokin was convinced that such perfect or Unlimited Love can best be explained by hypothesizing an inflow of love from a higher source that far exceeds that of human beings. Following the Russian tradition of integral knowledge, Sorokin was willing to hypothesize the existence of a higher source of love in the universe in which degrees of human participation are possible. He writes quite metaphysically of the exemplars of love at its fullest, many of whom were despised and had no psychosocial inflow of love to sustain them:

> The most probable hypothesis for them (and in a much slighter degree for a much larger group of smaller altruists and good neighbors) is that an inflow of love comes from an intangible, little-studied, possibly supraempirical source called "God," "the Godhead," "the Soul of the Universe," the "Heavenly Father," "Truth," and so on. Our growing knowledge of intra-atomic and cosmic ray energies has shown that the physico-chemical systems of energies are able to maintain themselves and replenish their systems for an indefinitely long time. If this is true of these "coarsest" energies, then the highest energy of love is likely to have this "self-replenishing" property to a still higher degree. We know next to nothing about the properties of love energy. Theoretically love may have its own "fission forces" that make its reservoir inexhaustible. When a person knows how to release these forces of love he can spend love energy lavishly without exhausting his reservoir.[18]

As evidence, Sorokin resorts to radical empiricism: that is, the legacy of human experience. Specifically, he refers to all the martyrs of love who, when surrounded by adversity, call out to a higher presence in the universe. Our understanding of such exemplars is very poor because science has not given them the attention they merit. Sorokin believed such enhanced understanding could increase love within individuals and society and between groups.

Sorokin's hopes that a scientific understanding of love might transform the world were grounded in human history and the "enormous power of creative love." He lists innumerable cases in which love stopped aggression and enmity, fostered love in turn, contributed to vitality and longevity, cured mental illness, sustained creativity in the individual and in social movements, and provided a basis for ethical life.

Sorokin openly asserted a view of human nature that included the *supraconscious*. This anthropological position underlies so much of *The Ways and Power of Love* that we cannot reasonably fail to consider it. Briefly stated, Sorokin speaks of the *biologically unconscious* aspect of the person, and assigns to it a somewhat diminished value:

Man is an animal, and all the reflexological, instinctive, and unconscious excitations and inhibitions, drives and activities of the
human organism necessary for animal life, growth, and survival
make up the lowest aspect of human personality—its biological
needs, drives, energies, and activities.[19]

Although Sorokin predated the emergence of ethology and evolutionary
psychology, much of the reflexive or instinctive rescue behavior (altruism)
that places the self at risk and that seems so deeply hard-wired into human
nature emerges from this biologically unconscious dimension of the self.
Although Sorokin takes this level of human nature seriously, his view of
love clearly emerges from higher cosmic energies rather than from the
biology below.

Related to this biologically unconscious aspect of the self is the *biocon-
scious in man*, by which Sorokin means all the solipsistic consciousness
related to the pursuit of biological needs and ego development. Thirdly, he
describes a *socioconscious in man*, including the "sociocultural egos, roles, and
activities" that we have as members of families, citizens of the state, occu-
pational and religious groups, and so on. He is keenly aware of how much
we are dominated by these social roles and personas, often to the point of
being overwhelmed. With a note of determinism, he describes the socioconscious as powerful and binding. This is relevant to his greatest concern, which is the foreboding specter of in-group insularity and intergroup
conflict in a technological world able to destroy itself.

Finally, Sorokin describes a "still higher level in the mental structure of
man, a still higher form of energies and activities, realized in varying
degrees by different persons—namely, the supraconscious level of energies
and activities."[20] This supraconscious level is explored in chapters six
through nine in *The Ways and Power of Love*; it is at this level that genuinely
creative love resides. Of course, Sorokin was running against the grain of
the social sciences, with their "materialistic and mechanistic metaphysics,"
and he therefore felt compelled to "lay down the very minimum of evidence" for the reality of the supraconscious. This evidence, as Sorokin
offers it, includes the *supraconscious intuition* that informs so much of the
highest human creativity (and the work of child prodigies) in virtually all
fields, from mathematics to ethics and religion. The *perfectly integrated creative
genius* achieves the highest level of creativity without strenuous effort. In

ego-centered love—that is, love "of low intensity, narrow extensivity, and short duration, impure and inadequate," no supraconscious is involved. However,

> quite different seems to be the situation with the supreme forms of creative love—intense, extensive, durable, pure, and adequate. Like supreme creativity in the field of truth or beauty, *supreme love can hardly be achieved without a direct participation of the supraconscious* and without the ego-transcending techniques of its awakening.[21]

Sorokin gathers empirical support for this statement from the testimony of "innumerable eminent apostles of love" who, across cultures and generations, describe themselves as instruments of the supraconscious: "God, Heaven, Heavenly Father, Tao, the Great Reason, the Oversoul, Brahma, Jen, Chit, the Supre-Essence, the Divine Nothing, the Divine Madness, the Logos, the Sophia, the Supreme Wisdom, the Inner Light."[22]

In part four, the longest section of *The Ways and Power of Love*, Sorokin sets out ideal typologies of loving and altruistic persons. He surveys the various techniques of altruistic transformation embedded in world religious traditions with an enormous scholarly depth. I suggest that this part of the book, entitled "Techniques of Altruistic Transformation of Persons and Groups," is by far the finest summary of such practices and rituals available.

The final part of the book, "Tragedy and Transcendence of Tribal Altruism," consists of a single chapter: "From Tribal Egoism to Universal Altruism." This is the last and most pessimistic chapter. Sorokin asserts a general law:

> *If unselfish love does not extend over the whole of mankind, if it is confined within one group—a given family, tribe, nation, race, religious denomination, political party, trade union, caste, social class or any part of humanity—such in-group altruism tends to generate an out-group antagonism. And the more intense and exclusive the in-group solidarity of its members, the more unavoidable are the clashes between the group and the rest of humanity.* Herein lies the tragedy of tribal altruism not extended over the whole of mankind or over everyone and all. An exclusive love of one's own group makes its members indifferent or even aggressive towards other groups and outsiders.[23]

Sorokin's concern with in-group insularity pervades his writings, especially in his many passages regarding the extent to which apostles of universal love have clashed with tribalists and been imprisoned, banished, tortured, and killed. In addition to exemplars of unlimited love for all humanity, innumerable groups have themselves been destroyed by the collective egoism of group loyalty. As Sorokin noted, "Whether in the form of a cold or a hot war, this intergroup warfare has gone on incessantly in human history, and has filled its annals with the most deadly, most bloody, and most shameful deeds of Homo sapiens." In-group exclusivism has "killed more human beings and destroyed more cities and villages than all the epidemics, hurricanes, storms, floods, earthquakes, and volcanic eruptions taken together. It has brought upon mankind more suffering than any other catastrophe."[24] Religious, ethnic, tribal, caste, and class wars have thus far defined much of human history and experience. What is needed, argues Sorokin, is enhanced *extensivity*. His recommendation is that the power of hatred be focused on threats to the whole of mankind, such as disease, ignorance, and poverty. He also recommends that competitions be sponsored on the basis of new values: "Unselfish love and humility can successfully be one of the most important competitive values." Indeed, *humility* was a core value in Sorokin's approach to a better human future.

In a visionary conclusion, Sorokin places his faith in science:

> Science can render an inestimable service to this task by inventory of the known and invention of the new effective techniques of altruistic ennoblement of individuals, social institutions, and culture. Our enormous ignorance of love's properties, of the efficient ways of its production, accumulation, and distribution, of the efficacious ways of moral transformation has been stressed many times in this work.[25]

Science can help us achieve the supreme good of "*sublime love, unbounded in its extensivity, maximal in its intensity, purity, duration and adequacy.*" It is certainly right to hope, with Sorokin, that progress in knowledge about love can move humanity forward to a better future.[26] In summary, what should we take from this discussion of the remarkable and controversial Sorokin? The message is that love and unlimited love can be captured by certain conceptual categories for potential measurement. In Sorokinian terms, then, "unlimited love"—a term he never used, but which fits his thinking well—is *love that is very high in intensity, extensivity, duration, purity, and adequacy.* Human love can become impressive in various of these aspects to significant degrees. The humanists among us, who perhaps eschew science and embrace only the subjective narrative of love, may scoff at the scientific approach. This summary of Sorokin's work may at least somewhat mitigate any such animosities.

3

THE CORE MEANING

OF "LOVE"

As we saw in the first two chapters, "love" is a word with many
meanings. In this consideration of unlimited love, our interest is
in love for neighbor as a pervasive affirmation of the very being of all
others, including self. This affirmation includes as background a thankfulness for the gift of life, a humility in the context of a fundamental
human equality, and a deep acceptance and patient tolerance of others
that is not thwarted by the inevitable imperfections, both internal and
external, in which we all share. It is an affirmation that leads us to take
interest in others, to be attentively present to them in a manner that is
undistracted and respectful, to be actively concerned with their well-
being, to listen to them with care, to be loyal to them in life's journey, to
act on their behalf with courage and fortitude, to delight in their suc-
cesses, and to require nothing in return. Depending on states of need,
love is appropriately manifest in compassion, forgiveness, service,
companionship, and a sage response to behavior that is destructive of
life and its potentialities.

All true virtue and meaningful spirituality is shaped by love, and
any spiritual transformation that is not a migration toward love is suspect. We are busy being reborn into lives of love, or else we are busy
losing ground. Many perceive that growth in love is a universal law of
life without which no meaning or lasting purpose is possible. It is also
widely perceived that the journey of love is pulled along by the alluring
nature of God, who assists us along the way.

A useful example of transformation in love emerges from the Russian tradition of the *ars moriendi* ("art of dying") literature, as represented

for modern readers by Leo Tolstoy's *The Death of Ivan Ilych*. Ivan is a remarkably unloving individual, callous and cold even toward family members; before his death, he realizes that his life has been all wrong. He rather suddenly feels that his schoolboy son has a life that is as meaningful as his own, or more so. This full emotional transformation away from solipsism toward the other enables Ivan to love the child.[1] Sometimes it is only the great equalizer, imminent death, that must be the mother of love.

PURE LOVE AND LOVE OF SELF

The love that interests us here is not predicated on reciprocal response. It makes no bargains and does not keep track of who reciprocates and who does not. It is pure in the sense of affirming the other as other, rather than for some ulterior motive. In reaching out to help others, new connections and relationships may unfold as a secondary effect, but these will have a uniquely unselfish tone because they are not entered into for personal gain. Deep community may emerge around helping behavior, but this must be distinguished from the self-interested, contractual relationships that only mimic true communion. Even if connections and relations do not emerge, the helping behavior continues unphased.

Moreover, the love that interests us here is not predicated on some internal growth or new level of well-being in the agent of love. No sense of enhanced self-esteem is directly sought. Yet paradoxically, in its effects such love usually contributes to the development of the self in profound ways, and will often enhance one's sense of purpose, meaning, and well-being.

Love for others includes caring for self with others in mind, and in the process discerning higher levels of dignity. "Love thy neighbor," and thereby discover the paradox of happiness in the forgetting of self. The self who has forgotten self-centeredness and lives close to Unlimited Love will take care of self, motivated not by self-interest but by *a totally different level of being*. Those who approach Unlimited Love will never be self-indulgent, yet they will be good stewards of their minds and bodies as instruments of love. Good self-stewardship remains obligatory, lest the agent of love undergo unnecessary emotional and physical deterioration, such that care for others becomes impossible. The agent of love is also the object of divine love, and has a responsive duty to tend to his or her dignity.

Self-stewardship requires us to pause and step away from the mode of constant "doing." Kirk Byron Jones, a theologian and Baptist minister at Andover-Newton Theological School, writes that we need to see value in simply being, rather than measure our worth through a calculus of activity: "Before you are a pastor, before you are a parent, spouse, or friend, you are a child of God, a person whom God loves unconditionally." In a useful distinction, Jones asks that we not "discard legitimate personhood along with the garbage of selfishness and egoism."[2] A great deal of the care of the self, argues Jones, involves taking the necessary time for silence, meditation, contemplation, and prayer.

To a large degree, true love of self is captured under the rubric of *inner peace*. The Dalai Lama, for instance, writes of spirituality as "those qualities of the human spirit—such as love and compassion, patience, tolerance, forgiveness, contentment, a sense of responsibility, a sense of harmony—which bring happiness to *both self and others*. While he speaks of a "radical reorientation away from our habitual preoccupation with self" that gives rise to helping behavior, this does not result in self-immolation but rather in self-discovery.[3]

I would hypothesize that living a life of love will, in the generalizable epidemiological sense, reduce morbidity and enhance longevity. Good care of the self for the sake of God and neighbor is probably more effective and enduring than care of self for the sake of self, which is less than fully meaningful. But longevity should not be counted on, and there will always be those who die young and well for loving purposes—think of Dietrich Bonhoeffer and Martin Luther King, Jr. We should have no illusions about the costs of a love that confronts malice. Many people discover inner peace and self-worth in loving others. The struggle to achieve self-esteem is often monumental and claims the lives of growing numbers of adolescents. Their lives are diminished, ruined, or lost in the cycle of *anomie* and self-destruction. The best way to achieve a sense of self-worth is through genuinely loving and serving others. As many admissions officers at outstanding schools will attest, the finest applicants typically have some significant experience in a voluntary association or faith community that teaches the joy of service to humanity. People who find meaning in love for others have found meaning for their lives, and they are thereby liberated from unnecessary suffering. They are released from the malaise of despair and the grip of peer pressure. They will discover new talents and abilities because they have something worthwhile around which to focus their energies. Paradoxically, then, those who lose themselves will find their truer selves (Luke 9:24). Let us "apply the whole measure of self love in love for neighbor," loving others with the same fervor with which we naturally love ourselves, thereby inverting self-love.[4] We will find again that there is a universal law: in the giving of self is the unsought discovery of self.

The description of love that I have provided thus far is consistent with that of the "radical love" discussed by the Jesuit Jules Toner, whose writings established an important school of thought in the last three decades of the twentieth century. His work *The Experience of Love* is considered a classic in the field.

Toner's description of the experience of love draws on the fullness of the human agent:

> In the full concrete experience of love, our whole being, spirit and flesh, is involved: cognitive acts, feelings and affection, freedom, bodily reaction—all these are influencing each other and all are continually fluctuating in such a way as to change the structure and intensity of the experience.[5]

By "cognition" Toner means memory, judgment, imagination, conceptualization, insight, and perception. By "affection" he means the experiences of joy and sorrow, love and hate, desire, hope, fear, and the like. By "freedom" Toner means the "power of self-determination by choice which is not determined by any condition or cause whether extrinsic to the agent or intrinsic to the agent but extrinsic to the act of choosing. It is the power by which a man can responsibly approve or disapprove, affirm or negate, his spontaneous affective responses."[6] By "bodily reactions" he means everything from heartbeat to facial expression. (It is important to keep this full picture of the agent in mind, for it suggests that love can and should be empirically studied with all these psychic and somatic aspects of the person in view.) In all of this concrete experience the self discovers itself.

Toner goes on to ask whether love is simply the composite of a number of affections, or whether it is unique. He sees the importance of empathic *care* for those in need, leading to the desire to assist altruistically, and appreciates why some will think of this as the root element of love. But such thinking is erroneous because love precedes and is wider than care: "And so, if love is basically caring or taking responsibility for someone, then it is never possible to love anyone unless I think he is in need. Nor is it even possible to love one whom I know is in need unless I am here and now considering his need." Love includes such things as a mother loving her child "in a moment of joyful security and careless playfulness," for love as "affection toward someone as a radical end without regard to need has a priority over care." In essence, "when care and need cease, love does not cease. If all need were to be fulfilled in the loved one, that would not mean the death of love. Care is only the form love takes when the lover is attentive to the beloved's need."[7] The object of love is the actuality of the person whether or not he or she is in need.

If radical love is not to be equated with care, what of the affection of *joy*? Toner sees joy as an essential constituent of love: "joy in the happy actualization of the one rejoiced over for and in himself."[8] Other-regarding love involves joy in a deep and broad sense. But love is not to be equated with joy, for joy is inappropriate when the other is suffering or grieving.

In the final analysis, argues Toner, "radical love" is "a response to the fundamental actuality of the beloved, to his [or her] radical act of personal being." Radical love does respond to lovable qualities and actions. Yet "love fails to be radical love of the other if in the other's qualities the lover fails to love the person for and in himself."[9] In a brilliant summary, Toner offers the following description of love:

> It is a response to the beloved's total reality. It is directly and explicitly a response to his actuality, fundamentally and in every instance to his fundamental actuality as a personal act of being; secondarily, to his qualitative actuality revealed in his acts and partially revealing his act of being. It is indirectly and implicitly a response to his potentiality, dynamism, and need. This response is experienced as liberation of the subject's energy for love and liberation from the confinement of individual being. It is at the same time experienced as a willing captivity to the beloved.[10]

Throughout, Toner emphasizes that love is a liberating experience for its agent, unlocking energy and creativity. Love for others sets us free from a myriad of limits.

EMPATHY AND COMPASSION

Love is not reducible to empathy, although empathy has a role in love. The innate and evolutionarily complex empathic capacity is an emotional feeling into the experience of the other that will often result in helping behavior. But the capacity to sense the experience of the other has no inherent moral or loving direction, and it is well recognized that some persons with developed empathic abilities may use them for manipulative or even nefarious purposes. Compassion, on the other hand, does clearly contain a morally beneficent direction, and can be understood as empathy linked with goodness.

Compassion is a vitally important modulation of love in the context of suffering. As Anne Harrington, a historian of science at Harvard University, defines compassion, "It is a process of external and internal reorientation that softens our sense of individuality by bringing it into a felt relationship with the pain and needs of some other."[11] Compassion requires empathy and seeks to achieve good in the context of suffering, and implies a readiness to be of help. Compassion is, therefore, moral in a sense that empathy is not; it affirms the goodness of the other as other, rather than as a means to further the interests of the agent.

Yet the very core of love includes a *grateful affirmation* of the other's very existence that has no particular correlation with suffering. Love is a radical affective affirmation of the other that is manifested by a desire to be with, and a willingness to participate in the life of, the other. This definition of love does not attempt to elevate compassion above all other manifestations. Instead, following more closely the insights of Buber and Levinas, I identify the core of love with an almost *sacramental appreciation and affirmation of the other*. This foundational disposition precedes any of love's other manifestations or modalities. Thus, love is to person as compassion is to person-in-suffering. Compassion means literally "to suffer with." It is always an impressive expression of love, and it is probably the case that in the context of evolution, compassionate love on the parent-child axis provided a beginning point. But unless we are all suffering all the time, love must take different forms and expressions than compassion. Miguel de Unamuno, in his classic work entitled *The Tragic Sense of Life*, mistakenly pictures all love (other than carnal) as a form of pity and compassion born of suffering and sorrow: "To love with the spirit is to pity, and he who pities most loves most." Unamuno continues:

Men aflame with a burning charity toward their neighbors are thus enkindled because they have touched the depth of their own misery, their own apparentiality, their own nothingness, and then, turning their newly opened eyes upon their fellows, they have seen that they also are miserable, apparential, condemned to nothingness, and they have pitied them and loved them.[12]

For Unamuno, love is compassion or it is absent. He laments the "vanity" and "tedium" of existence, while asserting that "all consciousness is consciousness of death and suffering."[13] Love has no other expression than compassion, he argues, because in reality there is nothing but suffering in life if we see it for what it is. He contends that God's love is also a matter of pity alone.

Unamuno was among the most graceful of the Spanish poets. His insights into suffering are profound. There is a pathetic aspect to our lives in that we do all suffer and fall short before the power of finitude. Yet Unamuno distorts human experience by omitting the *homo ludens* ("man the player") of play, celebration, and joy in the gift of life. Johan Huizinga coined this expression in his classic work describing play as an essential, indispensable mode of human existence. He moved far beyond play as essential only to the child. Huizinga saw play as a deep human reality upon which is based many of our highest spiritual accomplishments.[14] William A. Sadler, in an exhaustive phenomenology of love, also wrote that play is connected with love: "This is precisely what the world of love provides: love gives man a home in which it is safe to play."[15] Certainly love is contrary to fear and the inhibition of the playful nature of humanity.

Compassionate love is important and one of the most fundamental expressions of love, but love need not always be in compassionate form. (Indeed, those who always seem to wear compassion on their sleeves can be slightly annoying.) Love is associated with joy, play, and celebration in relational freedom. To cite Sadler, "Phenomenological investigations indicate that *the playground of freedom is love.*"[16] We should not in any way diminish the centrality of compassion or suggest that life is as playful as Huizinga suggests. But life is somewhat playful, and *play can be an expression of love in its joyful affirmation of, and participation in, the sacredness of the other.* Many aspects of religious ritual represent love in the modulation of play.

When the typical parent comes home, tired out after being at work all day, the first thing a young child wants to do is play; on this parent-child axis, joyful play with an inexhaustible child is surely a profound expression of love. The parent may have a gift for the child after a trip away from home. This also is an expression of love. Celebration and delight are key aspects of love that are clearly distinct from love's compassionate manifestations.

A great philosopher of love, Max Scheler, was correct in distinguishing the "big picture" of love from compassion, which he saw as "pity at its strongest." It is pity, of course, that, when genuine, "should lead to acts of beneficence." But love, Scheler argued, is too large a concept to be derived from compassion. He believed that such helping actions do not exhaust love. For Scheler, love is more a creative presence than a reaction to suffering or need. Love is expansive: "But love is a movement, passing from a lower value to a higher one, in which the higher value of the object or person suddenly flashes upon us; whereas hatred moves in the opposite direction." Love is about the vision of fullness and value in the other, and all the intentional acts that this implies. In the widest affirming and enhancing sense possible, love is a "creative force." In summary, "love is that movement wherein every concrete individual object that possesses value achieves the highest value compatible with its nature and ideal vocation; or wherein it attains the ideal state of value intrinsic to its nature. Hatred, on the other hand, is a movement in the opposite direction."[17]

Love, then, is not only manifest in compassion. Neither is it reducible to tender care in response to various need states, although we all recognize that tending to the needs of others is, like compassion, a core aspect of love. Love is to person as care is to person-in-need. The family or professional caregiver tends to the needs of the other, but the person may not always be in need. What then? Does love stop? The tendency in contemporary literature to speak almost exclusively of care rather than of love (except in love's romantic aspects) is remarkable, suggesting that we have lost sight of the higher meanings of love that give rise to the very existence of care and that are larger in scope than this highly significant manifestation.

We leave definitional distinctions and nuanced abstractions behind now and turn to some examples of love in practice, a transition which engages in the concrete and grounded realities of life and thus breathes greater life into what has been stated so far.

THE PRACTICE OF LOVE

Let us begin with Jean Vanier, who in 1964 founded the first l'Arche community for people with cognitive disabilities. Since that time l'Arche has grown into an international network of more than one hundred faith-based communities in thirty countries. Vanier's discussion of the nature of love is found in his book *Becoming Human*, which builds on his experiences in living with and serving persons with developmental deficits. Writing of his experiences in l'Arche, Vanier describes the rejection felt by persons with these disabilities. He writes of the importance of their feeling loved:

> In our l'Arche communities we experience that deep inner healing comes about mainly when people feel loved, when they have a sense of belonging. Our communities are essentially places where people can serve and create, and, most importantly, where they can love as well as be loved. The healing flows from relationships—it is not something automatic.[18]

Vanier describes Claudia, a seven-year-old autistic girl welcomed into a l'Arche community in Honduras:

> Her anguish seemed to increase terribly when she arrived in the community, probably because in leaving the asylum, she lost her reference points, as well as the structured existence that had given her a certain security. . . . She seemed totally mad, her personality appeared to be disintegrating. . . . Twenty years after she first arrived at Suyapa, I visited the community and met Claudia again; I found her quite well. She was by then a twenty-eight-year-old woman, still blind and autistic but at peace and able to do many things in the community. She still liked being alone but she was clearly not a lonely person. She would often sing to herself and there was a constant smile on her face.[19]

Vanier interprets her experience as a migration from loneliness and insecurity through community and love to inner peace. What does Vanier mean by love? Fortunately, he is specific about its seven aspects: *revelation of value; understanding; communication; celebration; empowerment; being in communion; and forgiveness.*

The first feature of love is the *revelation of value*: "Just as a mother and father reveal to their children that they have value and beauty, so, too, did the therapist and the others who lived with Claudia reveal to Claudia her value and beauty. To reveal someone's beauty is to reveal their value by giving them time, attention, and tenderness." In contrast to altruistic actions alone, "To love is not just to do something for them but to reveal to them their own uniqueness, to tell them that they are special and worthy of attention. We can express this revelation through our open and gentle presence, in the way we look and listen to a person, the way we speak to and care for someone."[20] The revelation of value can take time, but without it, Claudia's screaming madness was a viable response to a world that had rejected her.

I think the agent of love lives in an enchanted world where the other has sacred value. How can we become so enchanted, and can we see even the most downcast outsider in this bright Rembrandt-like ray of light? The revelation of value typifies *all* forms of love. Parental love reveals to the child his or her inestimable and unique value in a way that would seem impossible outside of parental investment and even infatuation. In conjugal love, each spouse reveals value to the other in ways that may, to some extent, be rooted in romantic infatuations; this value must be grounded more deeply in affirmation of being if it is to last. The great world religions all begin with the affirmation of value and meaning in all humanity, for all life is above all a gift over which we are stewards. Lovers of all humanity see unique and inestimable value even in the most devastated, imperiled, and seemingly unattractive human lives. It seems that the great lovers of humanity either find or bestow a special value in each person and can point this value out to the world. People want to be valued for what they are, not in spite of what they are not.

Such revelation inevitably provides the other with comfort, safety, and a release from anxiety. It is in stark contrast to the malignant social psychology that makes the other feel that his or her very being rests on a mistake; love is a revelatory affirming participation in the being of the other.

Vanier's second and third features of love are *understanding* and *communication*. Claudia "needed to be understood. If no one understood her how could they help Claudia to find inner peace and growth? Her screams were not only a sign of her inner brokenness, darkness, and anguish but also a cry for help."[21] Understanding precedes compassion and care, for the lover must become aware by understanding the other's suffering and need. Most people, including children, who feel unloved will lament that they were never listened to and therefore were never understood. Love manifests a readiness to understand the history of malignant devaluation. Empathy is certainly an aspect of this process of understanding, although such understanding involves a degree of intentionality and skill that is much more than empathy alone. Someone who loves must be communicative, a process that moves back and forth in an open, truthful flow of information, and that allows full and trusting articulation.

Vanier's fourth feature of love is *celebration*. "To love people," he writes, "is also to celebrate them." Love involves laughter, joy, and play: "The Claudias also need laughter and play, they need people who will celebrate life with them and manifest their joy of being with them."[22] Love relieves people from the perception that life is only a tragedy.

Love is not, in my view, best understood as a form of pity so much as an abrogation of rejection through the creation of a celebratory relationship. Negative self-esteem is finally conquered through celebration. It is the ultimate expression of inclusion, acceptance, and "being with." Celebration allows the fulfillment of the will to exist and the will to belong.

Vanier's fifth aspect of love is *empowerment*. As he writes of love, "It is not just a question of doing things for others but of helping them to do things for themselves, helping them to discover the meaning of their lives. To love means to empower." Claudia has to learn that she is responsible for her body, her life, and her choices. Further, "Empowerment meant that Claudia had to learn how to observe the structures of the community and make efforts to respect and love others."[23] Those who love do not attempt to possess or control or program, but rather wish to create community in empowered freedom. Empowerment allows one to have an identity based on a distinctive life journey, and to act in the light of that identity. An empowering love requires that the other be cared for with respect to needs, but it also wishes to teach the other to care for self; love will take the form of service as need demands, but it wants to encourage responsibility as well.

Vanier's definition of love evolves to include a sixth aspect, *to be in communion*. Through revelation, understanding, communication, celebration, and empowerment, Claudia is now able to participate in mutual trust and mutual belonging. Communion is the sustaining end point of love, "it is the to-and-fro movement of love between two people where each one gives and each one receives." Communion is not static, but constitutes an "ever-growing and ever-deepening reality that can turn sour if one person tries to possess the other, thus preventing growth." Vanier writes of openness to one another, growth in freedom, and, above all, of trust as essential to communion. In trust, "Claudia entered into a relationship of belonging. But we can only give of ourselves if we trust that we will be well-received by someone. At what moment is trust born? There was a secret moment, known only to Claudia, when she recognized that she was loved."[24]

In communion, the other has been discovered and accepted as worthy, and nurtured into trust and mutuality. The solipsistic tendency to view the other as valuable only insofar as he or she furthers "my" agenda has been completely set aside in all directions.

Vanier's final and seventh aspect of love is *forgiveness*. This aspect of love is central to communion, for we all need to be forgiven and to forgive. Forgiveness precludes the hatred and violence that destroy communion.

As we have seen, "love" is a remarkably complex term with a multitude of meanings. I believe Vanier has discovered its central meaning, and it is less a definition than a sketching out of aspects in chronological order: to reveal value; to understand; to communicate; to celebrate; to empower; to be in communion; to forgive. These are the observations of a man whose life has been devoted to creating a healing world of love for those imperiled by harsh rejection. The opposite of love is invalidation of being, and the related objectification, mockery, disparagement, and destruction of being.

We move now from Jean Vanier, whose ideas emerge from his experience as a person of unlimited love, to Tom Kitwood, known worldwide for his work with the most deeply forgetful—that is, with persons suffering from dementia. In his book *Dementia Reconsidered: The Person Comes First*, Kitwood writes, "I suggest that we might consider a cluster of needs in dementia, very closely connected, and functioning like some kind of cooperative. It might be said that there is only one all-encompassing need—for love."[25] Kitwood's cluster is resonant with Vanier's. He begins with *comfort*, which "carries meaning of tenderness, closeness, the soothing of pain and sorrow, the calming of anxiety, the feeling of security which comes from being close to another."[26] Kitwood then turns to *attachment*, which provides reassurance when bonds have been broken and the world is full of uncertainty. *Inclusion* follows, emphasizing the social nature of human life "related to the fact that we evolved as a species designed for life within face-to-face groups."[27] The need for inclusion comes to the fore in dementia because the loss of memory and communicative capacity have cut the person off from so many forms of community. Kitwood then refers to *occupation*, some way of drawing on a person's remaining abilities and powers to give him or her a sense of active agency. Finally, he includes *identity*, which flows from supportive community over time. Kitwood argues that love preserves personhood in the face of diminishing capacities. While Kitwood does not describe the chronological phenomenology of love as Vanier does, he too worked with a community of individuals who, however cognitively compromised, still needed to be loved and were often quite capable of returning love. (I remain convinced that we often are so busy that we forget about the centrality of love to human existence, and that persons with cognitive disabilities serve to remind us of our most basic human needs and nature.[28])

THE SCIENTIFIC TASK

In a time when cultural myths, images of human nature, and international conflicts may lead us to doubt the possibility of love and its manifestation in the world, we are inspired by those like Vanier who devote their lives to serving the neediest among us regardless of race, ethnicity, nationality, class, or cognitive condition. Telling the stories of extraordinary love is an important rejoinder to negative myths of human nature. Reflection on such narratives of helping behavior points us in the right direction.

Yet there is the equally important scientific response to pessimistic assertions. While science can not fully capture the hidden reality of Unlimited Love as an energy in the universe or clarify completely the extent to which human nature resists or accepts the ways of loving-kindness, science is central to our understanding of love just as it is to our understanding of the negative aspects of human nature to which we apply every scientific method available in our academic medical centers. We know a great deal scientifically about mental illness, about psychological deficits, and about abnormal psychology. Why not apply all these scientific methods to love, and related phenomena such as forgiveness, optimism, hope, and gratitude? If one were to make a small dot in the middle of a large white board, the dot would represent what we really know about love; the white represents what we don't know—which is a great deal. We proceed, then, with humility, and with the important qualification that in the end we need less to understand than to receive unlimited love.

Part Two

Scientific, Ethical, and

Religious Perspectives

4

LOVE AND THE SOCIAL SCIENCE

OF ALTRUISTIC MOTIVATIONS

" SCRATCH AN ALTRUIST and watch an egoist bleed." There are those who are convinced that no matter how genuine our love for others might appear, in reality human beings are only capable of selfish motivations. Scientists approach this question through research on altruism. Do we have altruistic motives? Do we act altruistically?

"Altruism" is a modern secular scientific concept whose sacred couterpart is *agape* love, although it lacks the emotional intonation of love. Both are other-regarding by definition, and imply self-giving without concern for response. It is difficult to imagine the invention of the word "altruism" without the background of a western culture in which the spirit of beneficence was broadened by the influence of *agape* or "charity." This background influence has been well described by the historians of western moral thought.[1] While altruism is, as we shall see, to some extent an inborn feature of human nature, culture and individual experience clearly influence the extent of its manifestation.

The hesitation of many theologians to engage in dialogue with the science of altruism is understandable because altruism emerged as a decidedly secular concept within the nineteenth-century domain of scientific positivism—that is, the view that science would eventually replace religion by substituting empirical reason for faith and superstition. The positivist view of history, of course, turned out to be false: the twentieth century clearly demonstrated that the influence of religion would rise rather than fall in a scientific world. The term "altruism," which derives from the Latin *alter* ("the other"), means literally "other-ism." It was coined by the nineteenth-century French sociologist and positivist Auguste Comte to displace terms burdened by a theological history[2] and was suggested by a French legal expression, *le bien d'autrui* ("the good of others"). However secular, Comte was nevertheless a child of European Christian culture; as such, he recognized the conflict between self-interested conduct and the morally appropriate domain of altruistic motives and behavior. He viewed the subordination of altruism to egoism as the source of human evil, and he was himself neither a psychological nor an ethical egoist. Because Comte envisioned a secular revision of *agape*, theologians have had little interest in dialogue with the sciences of altruism. But one need not endorse the secular humanistic tone of "altruism" to appreciate the scientific studies indicating the extent to which human nature manifests altruistic motives and behaviors. If there are any continuities or, perhaps better said, any points of correspondence or convergence between human nature and unlimited love, it is science that must describe the base material of our evolved propensities; this will require exploration of the forms of "altruism" in the human repertoire.

Is altruism a natural human possibility? If so, then Roman Catholics and Anglicans can find in human nature the hints of *agape* that suggest continuities between human nature and divine. If altruism is an ideal entirely contrary to our inborn egoistic nature, it is a concept that only serves to instill guilt and to verify the Protestants' pessimism that we are sinners incapable of the slightest good without the grace of God. In Catholic and Anglican thought we see a somewhat more optimistic view of human nature in which grace works not against nature but rather with nature to bring it to a higher level; on the other hand, Protestant thought views human nature as contrary to all true virtue. Theologians speak of the Anglo-Catholic synthesis of nature and grace as "both-and" (as in Thomas Aquinas), while the Protestant approach is described as "either-or" (as in Søren Kierkegaard).

For now, I wish to clarify that all claims of a hard-wired and intractable psychological egoism in human beings have been seriously discredited by the social sciences. This implies that Catholic and Anglican optimism may be warranted. Deep self-doubt about our helping motivations is generally unsubstantiated. Yet the more skeptical Protestant insight of Luther and Calvin remains useful in that it reminds us that ostensibly helping behaviors may be driven by self-interested motives that are usually observed in the form of self-righteousness, self-promotion, pride, and pomposity. Such observations, however, permit no wide generalizations about human nature. Often, altruistic motives are authentic and can be safely taken at face value.

Considerable scientific research, for example, indicates a direct link between compassion and helping behavior—that is, a link that skips the egoistic step of the agent asking, "What if that were me suffering?" and responds directly and in an unmediated way to the experience of the "other as Other." People seem to seek compassion in their suffering; in this sense, no matter how old we become, we are driven by a fundamental desire that is with us from birth.[3] Moreover, people have an innate propensity to be compassionate and to act helpfully, although this can clearly be culturally inhibited. If we are made in *imagine Dei* ("in the image of God"), then there must be some substrate of genuine love in our hearts.

That this substrate evolved under the genetic and environmental conditions of natural selection does nothing to detract from its authenticity. The influential field of evolutionary biology (once described as "sociobiology") finds genuine altruism not just on the parent-child axis, but increasingly at the level of group and community. While certain evolutionary biologists speak of genes as "selfish," such rhetoric need not be taken seriously. As Michael Ruse writes, "To talk of selfish genes is to talk metaphorically, and the whole point is that the phenotypes they promote are anything but selfish."[4] Genes are tools in the process of the continuation of life, rather than an underlying stratum of selfishness and greed that somehow tends to rise upward into actual living entities.

Human Nature in the Western Mind

The social sciences define altruism as a form of helping behavior that provides no anticipated material benefits to the agent and may incur some loss. Judith A. Howard and Jane Allyn Piliavin state that if purely psychological rewards are conceived as benefits to the agent, then virtually no human behavior would count as altruistic. Therefore, "many social psychologists maintain simply that altruistic behavior need exclude only the receipt of material benefits by the actor."[5] I almost agree with this position. I would prefer to assert that if an altruist retires by the fireside late at night and has some sense of meaning and fulfillment as the result of the day's helping behaviors, it seems counterintuitive to categorize him or her with the hedonic egoists of the world *so long as* he or she was not motivated by the sense of well-being that is retrospectively experienced. In other words, the sense of meaning and fulfillment must emerge paradoxically. To repeat a refrain found throughout this book, in the genuine giving of self lies the unsought discovery of a higher self.

Of course, if the altruist is motivated by the desire for material or reputational gain, he or she is not really an altruist.[6] Reputational gains may or may not attach themselves to the altruist, who may be confronting social injustice or in other ways violating society's prejudices against marginalized populations; the true altruist does not seek reputation, although this will usually accumulate and can be accepted as unavoidable. An altruist acts for the sake of others; the nonaltruist acts for the sake of self, or for self and others in some combination. Altruism can be combined with risk for the agent, but risk is not required. Altruists can be volunteer caregivers for the ill, builders for the homeless, and tutors for the illiterate without incurring significant risk. While those altruists who do "risk it all" in helping endeavors capture the public imagination with their intensity and selflessness, there is a great deal of unsung altruism, including the parental variant, that occurs in everyday life without significant risk. Altruism always requires a setting aside of the self as the center of the universe and a fundamental orientation toward others. Although altruism does sometimes demand significant risk, the agent merely needs to be open to it when it arrives.

There are individuals who exemplify helping behaviors that benefit others with no anticipated material benefit for the agent, that are universal in scope by embracing all humanity, and that can be sacrificial to whatever extent is necessary. But, as previously acknowledged, appearances can sometimes be deceiving. Many philosophers, theologians, and social scientists have been quite skeptical of all purported forms of even limited human altruism. Erving Goffman argued, in his classic work *The Presentation of the Self in Everyday Life,* that human beings are always performing before social observers. All the world really is a stage and we are all self-centered actors. We are taken in by our own performances and believe they convey our true altruistic nature. Self-deception makes one's altruism even more convincing, but no more genuine; those who are not taken in by their own routines are, therefore, less convincing.[7] A general trend in the social sciences had been to reduce all altruism to disguised egoism. But increasingly, when the current data are objectively considered, the presence of genuine (that is, psychological or motivational) altruism in human experience cannot be too seriously doubted, nor can the sense of a shared common humanity be dismissed.

If all altruism were nothing but disguised egoism, Luther would be vindicated. Some have asserted that we cannot escape controlling egoistic motivations, such as reputation, self-development, fame, health, longevity, wealth, vainglory, pleasure, status, and reciprocation. A civilization that emerged to some degree from the writings of St. Paul would hardly have an easy time asserting a genuine human altruistic concern for others due to the stark reality of sin (that is, a solipsistic inability to transcend self). Paul, Luther, Calvin, Jonathan Edwards, and many other theologians asserted the utter depravity of human nature; they maintained that the idea of neighborly love only indicts the moral conscience for its failures and thereby turns us to repentance, faith, and the graceful love of God. St. Paul put it well: "For I know that nothing good dwells within me, that is, in my flesh. I can will what is right, but I cannot do it" (Romans 7:18). Jonathan Edwards argued that there is no "true virtue" in human nature, and that participation in divine love frees us from all that is natural rather than elevates some substrate of natural goodness.

If altruism is nothing more than disguised egoism, the English Renaissance philosopher Thomas Hobbes is also vindicated. He was as pessimistic as Luther about human motivations, but offered no divine grace as a remedy. All he could offer was a set of prohibitions against wanton violence and harm in the name of enlightened or "long-term" individual self-interest, but the implementation of these restraints would require an all-powerful State (the Leviathan). Under the thin veneer of ostensible peace Hobbes saw an egoism controllable only by the power of fear. When questioned about giving alms to a poor man, Hobbes responded that his generosity allowed him to delight in seeing a poor man made happy, and was therefore a form of hedonism. Alasdair MacIntyre writes, "Thus Hobbes tries to exhibit his own behavior as consistent with his theory of motives, namely that human desires are such that they are all self-interested."[8] (Bernard Mandeville's *Fable of the Bees* was less an endorsement of Hobbes than a satirical account of the leadership of his times, which he perceived as hypocritical in its claims to benevolent motives.[9])

The tradition of extreme pessimism about what human nature brings to the table of authentic altruism moves beyond the writings of St. Paul, the Protestant reformers, and Hobbes. In modern times, one thinks of Sigmund Freud as the prominent psychological egoist, and of various misinterpretations of Charles Darwin such as Herbert Spencer's. Others, such as Friedrich Nietzsche and Ayn Rand, asserted that even if genuine altruism exists, we should immediately suppress it lest others become irresponsibly and weakly dependent. Anything else would be patently unethical to these "ethical egoists," who acknowledge the existence of genuine altruistic motives but wish to inhibit any altruistic actions in the name of an individualism that would have shocked the likes of Adam Smith. Smith, David Hume, and other moral philosophers in the Scottish Enlightenment insisted that genuinely altruistic motives do exist within the repertoire of human nature, and exhorted us to make the most of them.

Pauline theological thought and Hobbesian secular thought both recognized the ubiquity of human solipsism (from the Latin *solus*, alone, and *ipse*, self)—the tendency to view the self as the only significant entity. Others can be acknowledged and appreciated only insofar as they are in orbit around the self and are willing to contribute to its exclusively acquisitive agendas. Because human creatures are embodied and have various needs, they do often lack *ontological humility* in that they fail to appreciate the objective reality that no one person lies at the center of the universe, that others are equally worthy of existence and exist independently of the self's agendas.

Immanuel Kant offers an only slightly less pessimistic view than Hobbes. While he believed that certain altruistic helping inclinations exist, he urged us not to trust them because they are inherently unstable and therefore devoid of reliability. Moreover, he believed that inclinations have no moral value; duty alone has value as imposed by reason over all natural inclination, duty's antagonist. Altruistic behavior, which is possible for Kant, must be grounded in a categorical rational imperative.[10] The question for Kant was not the quality of desires, whether egoistic or altruistic by inclination, but rather what was required by human reason. He had no empirical interest in capacities other than reason alone. The power of autonomous reason, he argued, transcends inclinations, whatever they might be. Kant did assert an "imperfect duty" of general beneficence based on reason alone; by "imperfect" he meant that such a duty is flexible and more limited than duties not to do harm.[11] It is noteworthy that Kant never asserted the doctrine of psychological egoism, but was suspicious of other-regarding emotions as fickle. Reason alone, he argued, saves us from the shadows of malice, not love.

The Hobbesian skeptic, who sees in human nature only self-interested inclinations and a reasoning power that is an instrument of egoistic pursuits, does not deny the widespread existence of altruistic actions, only of altruistic motives. What better way to cater to pride, reputation, and self-glory than to behave altruistically? Among the strong Hobbesians are the remaining "classical" evolutionary biologists such as E. O. Wilson. The altruistic actor, of course, will genuinely and reliably give testimony to his or her other-regarding motivations—a phenomenon that contemporary evolutionary psychologists call *self-deception*. We are told that the illusion of goodness can be traced back to self-interest in the form of either social status or hedonic self-approval. The end result is a pleasurable pat on the back.

A full response to the Hobbesian skeptic—who is also the Pauline, Lutheran, Freudian, and Wilsonian skeptic—must be primarily empirical: Is human nature capable of genuine altruism, motivated by regard for others, or not? Science must step in, and much of eighteenth-century British moral philosophy and theology was heavily focused on observations about beneficent human action and its motivational sources. Many, but not all, philosophers arose to reassert the reality of "our amiable, that is, our moral affections, as analogous to God's."[12]

The social sciences continued these classical discussions into the modern period. The scientific and the perennial philosophical discussions are quite continuous: they ask the same question about human nature and base their answers on careful observations of human action.

Contemporary Social Science Research on Altruistic Motivations

C. Daniel Batson, the premier contemporary social scientist concerned with altruism, defines the term as follows: "Altruism is a motivational state with the ultimate goal of increasing another's welfare."[13] Batson adds that altruism does not necessarily involve self-sacrifice, although it seems to me that it is inherently self-sacrificial to some degree, and that this should be acknowledged. The key feature of his definition is the genuine motivational state that seeks the good of the other: "Egoism is a motivational state with the ultimate goal of increasing one's own welfare."[14] As Batson defines an essential question regarding altruism, "We want to know if our helping is always and exclusively motivated by the prospect of some benefit for ourselves, however subtle." He continues:

> The question thus raised has been a central one in moral philosophy for many centuries; it is the question of the existence of altruism. Advocates of universal egoism, who are in the clear majority in Western philosophy and psychology, claim that everything we do, no matter how noble and beneficial to others, is really directed toward the ultimate goal of self-benefit. Advocates of altruism do not deny that the motivation for much of what we do, including much of what we do for others, is egoistic. But they claim that there is more. They claim that at least some of us, to some degree, under some circumstances, are capable of a qualitatively different form of motivation, motivation with an ultimate goal of benefiting someone else.[15]

The practical stakes in answering this question are high: If we are only capable of seeking self-benefit, then all talk of altruism is nonsense; ethics is reduced to contractarian negotiation among various competing self-interests. Most scholars consider ethics to be the study of how we move beyond egoism to take into account the needs of others. The contractarian school of ethics, best represented in modern times by John Rawls, does take into account the long-term or enlightened self-interest on the part of "rational and mutually disinterested" agents, but this is a very limited and weak approach that falls short of a true commitment to others as others.[16] If altruism is motivationally implausible, then pedagogical efforts to create a more ethical society must be appropriately designed along contractarian lines. But if we can reach beyond egoism based on natural inclination, ethics becomes genuine rather than a contrived contractual endeavor to restrain the wolf within. As Batson puts it, "To build a better society, we need to know the truth about our raw materials."[17]

It should be clearly understood that the altruism question is not about the existence of helping behaviors. External helping actions are commonplace. The question is whether such actions are ever really controlled by other-regarding motives. Altruistic conduct can be controlled by the need to reduce the aversion caused by seeing people suffering; in this case, helping removes unpleasant feelings. Such conduct can also be controlled by *internal* rewards to self, such as self-satisfaction or self-esteem. A Freudian model of the super-ego, for example, would be entirely consistent with altruistic conduct that is driven by the need for self-satisfaction. Altruistic conduct can also be shaped by a controlling interest in *external* rewards to self, such as reciprocity or social recognition and success. Batson refers to egoistically controlled altruistic conduct as *"pseudo-altruism."*[18]

According to Batson, altruistic motivation must involve the empathy-altruism hypothesis, in which the agent feels the needs of the other on an emotional level and engages in helping behavior. He seems to be somewhat dismissive of the Kantian rationality-altruism hypothesis, in which a reasoned understanding of the equal claims of others leads to helping behavior; he suggests that reason alone is not obviously a spur to action. (By "empathy" Batson seems to mean compassion.)

As Batson frames the discussion, there are two egoistic accounts of why we engage in helping behavior, one of which (Path 1) "is based on social learning and reinforcement; the other [Path 2] on arousal reduction."[19] Path 1, social learning and reinforcement, is divided into reward-seeking (being paid, gaining social approval, receiving reciprocated help) and punishment-avoiding motives (avoiding censure, avoiding guilt). Path 2 is independent of anticipated rewards and punishments. Seeing the other in distress may cause the agent to feel disturbed, anxious, and upset, to the extent that helping behavior follows as a matter of self-interested relief. Batson argues that paths 1 and 2 are not mutually exclusive. Both involve a "hedonic calculus," even if this must be rapid in some circumstances.[20] Paths 1 and 2 are both forms of pseudo-altruism. While there are many social-scientific studies that support the egoistic pathways, numerous researchers also argue for a third path of true altruism, in which empathic emotion evokes altruistic motivation.

Path 3, the empathic pathway, requires the agent to adopt the perspective of the other, imagining how the other is being affected. The Greek *empatheia*, "feeling into," refers to the capacity to feel the subjective experience of the other. The agent does not ask what he or she would experience under these circumstances, but what the other, as other, is experiencing.

Taking up the perspective of the other is generally preceded by attachment to the other.

In essence, empathy requires an intuiting of what the other is experiencing, which requires both a cognitive or imaginative dimension as well as an other-oriented emotion that is variously called "compassion," "pity," or other such terms. Batson claims that the prototype for the attachment that underlies empathy is "the parent's attachment to the child."[21] But such attachments encompass other family relationships and friendships. Batson uses the term "attachment" in a way that is more or less equivalent to "love." The strength of the attachment, coupled with the magnitude of the perceived need, determines the strength of the empathic emotion. With regard to truly altruistic conduct:

> According to the empathy-altruistic hypothesis, magnitude of the altruistic motivation evoked by empathy is a direct function of magnitude of the empathic emotion. The more empathy felt for a person in need, the more altruistic motivation to have that need reduced. Reducing the need of a person for whom one feels empathy is likely to enable the helper to gain social and self-rewards (Path 1a), avoid social and self-punishments (Path 1b), and reduce feelings of personal distress (Path 2). The empathy-altruism hypothesis claims, however, that feeling empathy for the person in need evokes motivation to help in which these benefits to self are not the ultimate goal of helping; they are unintended consequences.[22]

In other words, empathic altruism is genuinely other-regarding. This empathy-altruism hypothesis is historically prominent in the writings of Adam Smith, David Hume, Charles Darwin, and William McDougall, and has considerable support among contemporary researchers.

There are, however, serious empirical studies supporting the view that the empathically aroused individual is, in fact, motivated to diminish or turn off this arousal; therefore, even empathic altruism can be described within the egoistic boundaries of Path 2 in that it alleviates distress. After extensive review of the available studies, Batson concludes that, contrary to the Path 2 explanation, empathic helping "does not pattern as we would expect if the goal were aversion-arousal reduction." Nor does the evidence suggest that empathic altruism is ultimately motivated by the desire for rewards or the fear of punishments.[23] Batson states, "In study after study, with no clear exceptions, we find results conforming to the pattern predicted by the empathy-altruism hypothesis, the hypothesis that empathic emotion evokes altruistic motivation. At present, there is no plausible egoistic explanation for the results of these studies." He therefore urges that the empathy-altruism hypothesis be tentatively accepted as an answer to the altruism question: "Contrary to the beliefs of Hobbes, La Rouchefoucauld, Mandeville, and virtually all psychologists, altruistic concern for the welfare of others is within the human repertoire."[24]

While I find Batson's work persuasive, I would add that the empathy-altruism axis does need the enhancing and stabilizing power of reason and the spiritual perspective of a common humanity if human altruism is to extend consistently beyond the narrow domain of the near and dear into that of all humanity and the neediest. Ever since Plato, who emphasized the primacy of reason in moral motivation, ethics has debated the proper role of emotion in the moral life. Adam Smith, Charles Darwin, and Batson define an empirical tradition that makes other-regarding emotion central. Using Batson's definition, all three assert the importance of empathy, the "tender emotion." Smith uses the word "sympathy," but he virtually equates it with "empathy" in the contemporary context. While I do not wish to depart from this empirical tradition and assert the primacy of reason, I do think that reason must be more than merely instrumental to altruistic inclinations. An even balance or co-primacy between emotion and reason is the fitting alternative to those who would diminish the importance of either capacity. Reason is given a primacy in both utilitarianism and Kantian ethics, making emotion an irrelevant obstacle to the moral life. This is regrettable. I defend Darwin, who developed the creative synthesis of co-primacy, recognizing the altruistic powers of both emotion and reason. He finds "that short but imperious word *ought*" to be "full of high significance" as he extols Kant in particular for exclaiming to the world, "Duty!"[25] Darwin saw reason as a source of motivating authority that is largely free from genetic determinism and able to work in consort with the moral emotions as it recognized the principle of duty to all humanity in a kingdom of ends.

Batson's empathy-altruism axis is a natural building block of love. But the loyalty and attentiveness of love are not things that lie dormant within us. Moreover, empathy can be used for evil purposes. Our natural tendency toward helping behavior, which is quite instinctive by Batson's model, can be enhanced by cultural influences, the influence of exemplary individuals, reason, and spirituality in order to be fully elevated into something higher. One weakness of the empathy-altruism hypothesis generally is its relative diminution of the role of these other factors in the growth of compassionate love. Kristen Renwick Monroe, for example, argues that "the perspective-taking aspect of empathy in itself is not sufficient to cause altruism, since this increased understanding need not necessarily be utilized for the other person's welfare."[26] Monroe recognizes the established importance of empathy, but couples it with the altruist's perception of a shared humanity: "The idea of being welded together, of belonging to one human family, surfaced over and over again in my interviews; indeed, I was struck by the similarity of expression used, particularly since I myself was careful to avoid terms such as 'one family of man' and never suggested such a view in my questions."[27]

Altruism flows from a sense of self-identity that emerges from a perception of self as part of "all humanity." This is true, at least, for altruism that is fully inclusive. Norms and beliefs, worldview, and view of self (self-identity) inform the empathic capacity to create altruism.[28] Whether this "self-identity" model of altruism is significantly different from the Kantian one is questionable; it may simply be that the Kantian principle of universal respect for humanity is captured in some form within the particular tradition or profession with which the agent of altruism identifies.

In summarizing the above material, I think it is fair to state that many social scientists and philosophers do discern in human nature an empathic capacity that responds to the circumstances of others in a direct and unmediated manner, and which results in altruistic actions. It is probably true that, at some level, this capacity cannot be entirely inhibited in most of us, no matter how much we are acculturated to images of self-interested individualism. In times of catastrophe, when the routines and illusions of everyday life are upset, this capacity will often rise powerfully to the surface, resulting in a great deal of spontaneous helping behavior.

While Batson's work is generally well received, his reliance on the word "empathy" raises the question of whether "compassion" would be the better term. Empathy is really an epistemological experience of "knowing" the feelings of the other; in contrast to compassion, it lacks any obvious moral direction. Batson uses the term "empathy" in a general way to refer to a "set of congruent vicarious emotions, those that are more other-focused than self-focused, including feelings of sympathy, compassion, tenderness, and the like," which he distinguishes from feelings of personal distress and equates with pity (Aquinas, Hume, Smith), compassion (Hume, Smith), or tenderness (McDougall).[29] Batson opts for "empathy" because, he suggests, the other terms are either "moralistic" or linked with emotional contagion.

The philosopher Martha C. Nussbaum, however, makes the case for using the word "compassion." In her extensive analysis of linguistic distinctions, she begins with a general bifurcation of the emotions into those that expand moral boundaries, such as compassion, and those that draw sharp boundaries, such as disgust or egocentric forms of love. She prefers to use the word "empathy" in a way that clearly distinguishes it from "compassion." Empathy is the "imaginative reconstruction of another person's experience, without any particular evaluation of that experience." Thus, "a malevolent person who imagines the situation of another and takes pleasure in her distress may be empathetic, but will surely not be judged sympathetic. Sympathy, like compassion, includes a judgment that the other person's distress is bad."[30] One might use empathy for egoistic and evil purposes, as Hitler did: Does empathy contribute anything of ethical importance entirely on its own (when it does not lead to compassion)? I have suggested that it does not: a torturer can use it for hostile and sadistic ends.[31] Nussbaum equates the emotions of sympathy and compassion because they involve a structure in which the agent understands the threat to the flourishing of the other as serious and not due to his or her fault; the agent is not driven by the prudent fear of being in a similar situation. She argues persuasively that compassion, unlike empathy, is clearly linked to beneficent actions; thus, she draws on Batson's studies, but only after transposing his language appropriately. Batson's work convinces her that compassion leads to helping behavior because the suffering of a separate being is bad in itself, and not because of a feeling of one's own vulnerability ("There but for the grace of God go I"). As Nussbaum describes it, the experience of compassion is an emotional "upheaval"—borrowing this term from Marcel Proust—and in this sense is painful. Yet scientific studies do not permit us to reduce the helping actions that flow from compassion to the hedonic relief of the agent.

A COMMENT ON THE RELIGION FACTOR

While I appreciate Batson's clear demonstration of the reality of motivational or psychological altruism, I am critical of his dismissal of the evidence that religion does play a role in encouraging altruistic behavior.[32] In an early study, Nelson and Dynes explored the impact of religious devotion and attendance at religious services on a variety of helping behaviors.[33] The researchers mailed a questionnaire to a sample of adult male residents in a city that had recently been struck by a tornado. The subjects were asked three questions to determine their level of devotion: 1) How often are table prayers said at mealtimes in your home? 2) How often do you pray privately or only with your wife (excluding mealtimes at your home)? 3) How important is prayer in your life? Respondents were also asked how frequently they attended church (at least weekly, one to three times monthly, less than monthly, or never) and to rate their level of religious commitment (deeply, moderately, not very, or not at all religious). The subjects' responses to these questions were then compared with their involvement in both "ordinary" helping behaviors (contributing food or goods to social service agencies, helping motorists with car trouble, picking up hitchhikers, participating in volunteer work, etc.) and emergency helping behaviors (donating money to relief organizations, providing relief goods for tornado victims, and performing disaster relief services unrelated to regular employment). The researchers found that devotionalism, church attendance, and level of religious commitment were positively correlated with levels of helping behavior, both in routine and emergency situations.

In a more recent study, Hunsberger and Platonow examined a group of nearly four hundred students to investigate the links between spiritual and religious beliefs and practices and charitable behaviors.[34] The researchers found that religious students were more likely to report volunteering and donating money to church-affiliated activities, but found little difference between religious and non-religious students regarding volunteer activities for "secular" charitable organizations. Those with an intrinsic religious motivation (religion practiced for its own sake and not for another more external purpose, such as social desirability) were more likely to volunteer than those with an extrinsic religious motivation. Most intriguingly, intrinsic religiousness best predicted overall volunteer practices and attitudes. Somewhat unexpectedly, intrinsic religious commitment was also highly correlated with orthodox religious beliefs—beliefs often viewed by the public as being associated with those least willing to help a hurting society.

F. M. Bernt also found a link between intrinsic religiousness and altruistic behavior. College students were asked how many hours they volunteered during a semester, and were also asked to answer questions from intrinsic and extrinsic religiousness scales.[35] The researcher found that students who volunteered for service activity had a more intrinsic and questing type of faith than those students who tended not to volunteer. The link between volunteerism and intrinsic religiousness is an important one, as it may highlight the deeply personal and internalized spiritual beliefs that possibly motivate altruistic behavior.

Dr. Jan Shipps, professor emeritus of religious studies, history, and philanthropic studies at Indiana University-Purdue University in Indianapolis, explored how the institutional configuration of religion in cities can affect giving practices. When examining data reported to the Internal Revenue Service for seventy-two midsize cities, she concluded that religion and philanthropic giving are closely linked. "The percentage of church adherents in a city's population is a good predictor of the amount of total itemized charitable deductions that people who live in that city will report to the IRS."[36] Indeed, a recent Yankelovich survey of American giving trends found that the most significant predictor of large donations by individuals is weekly attendance at religious services.

These findings seem consistent with spirituality and religion at their best in that they illustrate the appropriate ordering of emotions and the elevation of positive emotions over negative ones. The famous Prayer of St. Francis, for example, is chiefly a spiritual exercise to enhance love over hate:

> Lord, make me an instrument of your peace. Where there is hatred, let me sow love; where there is injury, pardon; where there is doubt, faith; where there is despair, hope; where there is sadness, joy; where there is darkness, light.

Likewise, the American Puritan theologian Jonathan Edwards, in his classic treatment of the "religious affections," highlighted the essential dominance of emotions such as love and joy in any genuine spiritual life.[37] William James, in his *Varieties of Religious Experience*, concluded that a shift toward warm other-affirming affections is the key feature of significant spiritual experience across traditions and epochs.[38]

Within the monotheistic traditions, the spirituality of love and compassion is supported by the perception of a God whose love is steadfast and whose mercies are tender. Compassionate love is connected with the emotions of gratitude or awe before the wondrous works of God in the created universe and the emergence of life. It is connected with the emotions of hope and faith before the loving purposes of God that are never extinguished and will eventually be fully realized. It is connected with the emotions of joy and peace before the presence of the Lord, and with the emotions of sorrow and repentance before a God who manifests perfect love and justice. It is connected with the emotions of confidence and courage before the knowledge that loving purposes are God's own. It is connected with the emotion of forgiveness before a God who is the most forgiving of all, and with the emotion of humility before God's infinite love. The emotions of contempt, disgust, and hatred are simply contrary to the nature of a God of love. Within a spiritual context, to behave with compassionate love is to be in resonance with one's maker. Thus, compassionate love is enhanced in high spirituality because it is supported by a cluster of other emotions and is afforded the very highest status, for in love we act in the image of God and in ultimate human dignity.

CONCLUSIONS

Everyone has faith in something of ultimate concern that pulls them in one direction rather than another. Faith in the broadest sense need not be religious in any way. Any institution or purposeful life is predicated on loyalty to or faith in some center of value. In the words of James W. Fowler, faith is in "an image of the ultimate environment. . . . If the term ultimate environment gets in the way, let us speak of a comprehensive frame of meaning that both holds and grows out of the most transcendent centers of value and power to which our faith gives allegiance."[39] Faith in Unlimited Love as the "ultimate environment" is something different from faith in a rational precept or principle. In the person of love every human capacity is in tune and the enduring emotion of other-regarding kindness can be felt and heard like splendid music.[40] Love is neither just a rational principle, a vision, a symbol, or a vision, although it is all these things; it is in essence a deep and stable attunement of the emotional self to the well-being of others that can be refined and elevated to remarkable degrees in the life of every person. On some level, I think that we all want to place our faith in Unlimited Love as the "ultimate environment" of the universe, for it is frankly difficult to imagine anything less than this love providing integration, meaning, and purpose in our lives. Love inspires, illumines, and leads the way in the lives we most admire. In the words of the prophet Micah quoted earlier, "What does the LORD require of you but to do justice, and to love kindness, and to walk humbly with your God?" (6:8). In my view this is, simply put, how we should live.

In this chapter I have presented a scientifically plausible image of natural altruistic motivation toward humanity. In essence, however much selfishness exists in our nature, there is at least an equal and opposite unselfishness. Even in his pessimism, St. Paul recognized this when he wrote in his letter to the Romans: "For I delight in the law of God in my inmost self, but I see in my members another law at war with the law of my mind, making me captive to the law of sin that dwells in my members" (7:22–23).

5

The Evolution

of Altruistic Love

O NE APPROACH to the question of the existence of genuine human
motivational altruism is that of evolutionary biology and the closely
related evolutionary psychology. There is reason to be more impressed by
the social scientific research of Batson, which directly attends to human
behavior, than by the extrapolation to humans from nonhuman species
and game theory that is typical of evolutionary biology. But evolutionary
thought has much to contribute to an understanding of the extent to
which human nature is a welcoming substrate for unlimited love.

Evolutionary biologists contend that humans possess a narrow altruism toward genetically related others ("kin"). The narrow scope is not
based on any careful descriptive studies of how humans actually behave
or report motivations. Instead, the narrowness emerges from an explana-
tory model of genetic determinism that is derived from mathematical
models and lower species lacking our cultural malleability, freedom of
will, and moral creativity. Richard Dawkins, for example, in *The Selfish Gene*,
writes:

> The argument of this book is that we, and all other animals, are
> machines created by our genes. Like successful Chicago gangsters,
> our genes have survived, in some cases for millions of years, in a
> highly competitive world. This entitles us to expect certain qualities
> in our genes. I shall argue that a predominant quality to be expected
> in a successful gene is ruthless selfishness. This gene selfishness
> will usually give rise to selfishness in individual behavior. However,
> as we shall see, there are special circumstances in which a gene can
> achieve its own selfish goals best by fostering a limited form of

altruism at the level of individual animals. "Special" and "limited" are important words in the last sentence. Much as we might wish to believe otherwise, universal love and the welfare of the species as a whole are concepts that simply do not make evolutionary sense.[1]

The overall impression given by such rhetoric is that human nature is fully determined by so-called "selfish" genes that are somehow motivationally controlling. Of course genes are merely bits of DNA, incapable of motivations themselves, and as elements in human nature do not explain the human being in any full sense. Dawkins uses "selfish" metaphorically, but in so doing mischievously paints a portrait of human nature that is likewise selfish. Dawkins allows for a very "limited" potential for altruistic motives and behavior, but his "robot survival machine" can only be altruistic to the extent that this serves the purposes of successful procreation, whereby genes migrate forward into the next generation of survival machines. This reductionism is philosophically indefensible.

Contrary to Dawkins's strong genetic determinism, even the most casual perusal of human behavior indicates that a new phase of creativity and freedom has emerged to provide a novel level of causality "from above," involving culture, mind, moral sense, and spirituality. These directive powers cannot be reduced to "nothing but" the lower substrate of genes, anymore than the writing of a splendid symphony can be understood simply in terms of pen, ink, and paper. There is a causality within the novel emergent capacities of the human that can redirect and profoundly expand altruistic propensities that originally evolved for narrow purposes.[2]

This is not to suggest that evolutionary biology fails in explaining the evolutionary beginnings of human altruism in nonhuman species, but only that it misses the ways in which human altruism moves far afield from restrictive beginnings on the procreative axis. This expansion occurs through a variety of influences, including the survival needs of the larger group (group level selection), the influence of culture, the moral logic of human equality, the influence of religious and ethical wisdom, and, as many would add, the direct power of unlimited love. The evolutionary biologists may succeed in explaining the beginnings of altruism, but seem almost oblivious to the full narrative of human experience. One wishes that they might take more seriously the course of human events, such as Christian parents in Nazi Germany rescuing Jewish children and caring for them as their own, all the while knowing that should their altruism be discovered, the Nazis would eliminate them and their biological children.

EARLY-SCHOOL EVOLUTIONARY BIOLOGY

In essence, Dawkins writes within the framework of early (mid-1960s) evolutionary biology and evolutionary psychology, for which the goal of all behavior is reproductive success, which means passing on one's genes. This genetic preservation, however, can be accomplished to varying degrees by sacrificing personal reproductive success to the success of others who carry one's own genotype. Behavior that benefits those who are genetic kin—most often offspring—is altruistic, and is described as "kin-altruism" or "inclusive selection." All other behavior is a matter of enlightened self-interest that contributes to core reproductive success. Thus, in nonkin relationships, the framework is one of self-interested reciprocity and cooperation to enhance procreative goals. (For example, one tends to be quite cooperative at work and engage in all the expected social reciprocities because this allows one to support the needs of one's family.) Acts that may seem genuinely altruistic within this cooperative nonkin context are only ways to enhance one's long-term social reputation as a reciprocator; when we think that we are genuinely altruistic outside of kinship, we are fooling ourselves. Such self-deception has its evolutionary advantages because it allows the agent to gain credibility, with consequent additional benefits. Social relations outside of kin-altruism are never as "good" as they might appear, for their base metal is strategic egoism. Humans evolved brains that are sophisticated scorekeepers: this one is a good reciprocator, this one is a "cheater," this one deserves our "grudges." Or so we are instructed by early-school evolutionary biology. Much of the early-school emphasis on self-interested reciprocity is theoretically rooted in mathematical game theories that assume motivational selfishness. The game theories underlying this biology are themselves cultural products of economic individualism.

Contrary to early evolutionary biology, human altruism does move beyond the narrow confines of kin-altruism, spilling over into larger groups and, with the help of culture, reason, and spirituality, into the domain of love for all humanity. This expansion of kin-altruistic emotions is appealed to every time a stranger on the street says, "Brother, can you spare a dime?" As Catherine A. Salmon's studies of subjects listening to kin speech underscores, the use of such speech clearly does elicit the emotions associated with close kin ties, and results in increased altruism. In politics and religion, the natural affective solidarity between kin is called forth by manipulation of kin metaphors, and thus "we have all heard speakers who wish to promote beneficence address nonrelatives with kin terminology."[3] This expansion is measurable both physiologically and behaviorally. Early-school evolutionary biology does not accept this notion of genuinely altruistic spillover into larger groups, and on this point we must embrace the newer developments in the field.

New-school evolutionary biology argues for group-level selection and genuine altruism toward nonkin. In other words, wider social relations are not constrained by "Tit for Tat" reciprocities along the lines of game theory. The driving force in group selection and group altruism is not reciprocity, but something much more like kinship altruism writ large. The good news here is that we can care for others in the group more or less as we do for our own kin; the bad news is that the scope of this altruism can be limited by group interest, suggesting conflict between groups as the cost of in-group solidarity. New-school thinkers tend to be fatalistic when it comes to intergroup conflict and are skeptical of universal benevolence. The world of current events seems filled with intergroup conflict along ethnic, racial, class, and nationalistic lines. Mutually acceptable treaties keep a fragile peace.

Co-evolutionary theory provides a third framework, different from both early and new-school evolutionary biology. It contends that human behavior breaks free of the restrictive leash of genetic determinism through the influence of culture. A cultural or religious tradition that teaches love for all humanity might modify the base metal of human nature into something greater. Religious ideas and rituals, reasoning powers, and various pedagogical influences can create a new magnitude of altruism that builds on the authenticity of kin-altruism and group altruism to the love of all humanity without exception.

The remarkable narrowness of early-school evolutionary biology is evident in this passage from George C. Williams's 1966 study, *Adaptation and Natural Selection*, in which he wrote: "The natural selection of alternative alleles can foster the production of individuals willing to sacrifice their lives for their offspring, but never for mere friends."[4] A clear factor in evolution, he continued, is the advantage that accrues to those who maximize friendships and minimize antagonisms. Coalitions of mutual aid are commonplace, with their origins strictly in the domain of self-interest. Human beings abide by the rule of nature—there is safety and enhanced survival for the individual in the group, but the individual is always the self-interested unit of selection. By analogy, "The huddling behavior of a mouse in cold weather is designed to minimize its own heat loss, not that of the group."[5] Williams based this assertion not on observation of human behavior, but on a methodological assumption that when biological events can possibly be explained on the basis of individual genetic self-interest, such a theory should hold sway over alternative "groupish" interpretations. Yet contrary to Williams, people do give up their lives for their friends, and only torturously strained theoretical assumptions can cast unconvincing doubt on this human reality

GAME THEORY: A DIMINISHED IMAGE OF HUMAN NATURE

The work of Williams and others was supported in 1984, when Robert Axelrod asked, "Under what conditions will cooperation emerge in a world of egoists without central authority?" People "are not angels, and they tend to look after themselves and their own first."[6] So how can cooperation develop in a universe of selfish agents? The winning strategy was Anatol Rapoport's "Tit for Tat," which indicates that an agent should cooperate until it meets with a defection (a nonreciprocator or "cheater"), at which point it must punish (bear grudges, shun, exclude). In other words, cooperate on the first move and then do exactly as the other does (Do unto others as they have done unto you!). This is a conceptual model lacking in any forgiveness, of course, and one that requires a great deal of careful scorekeeping. It is difficult to imagine that any human being would wish to live in such a system, which would quickly devolve into a world nicely portrayed in the movie *The Unforgiven*. Axelrod, in a chapter entitled "The Evolution of Cooperation in Biological Systems," co-authored with early-school evolutionary biologist William D. Hamilton, states that almost all cases of altruism in nature exist in the context of "high relatedness, usually between immediate family members." Furthermore, "True altruism can evolve when the conditions of cost, benefit, and relatedness yield net

gains for the altruism-causing genes that are resident in the related individuals."[7] Outside of such genetic relatedness, altruism is replaced by
cooperative strategies. The influence of "Tit for Tat" on early-school evolutionary biology is so significant that we must pause for reflection. In
playing a game, the agent follows the rules and goals of maximum rewards
("optimality") by making the best decisions. Game theory offers a comprehensive model of decision-making in mathematical terms. Through the
seventeenth century, various thinkers developed mathematical models considered applicable to human experience. Leibnitz, for example, recognized
in the study of games the combination of chance and skill representative
of human life. In the early twentieth century, major mathematicians studied games, leading to the important work of John von Neumann. Von
Neumann and Morgenstern established the field as a foundation for economics with the 1944 publication of their classic *The Theory of Games and
Economic Behavior*. The language of "utility," "payoffs," and "optimal strategy" dominates game theory, and thus colors all the areas of human activity to which these models are applied, including politics, warfare,
economics, and evolutionary biology.[8] Scholarship in these areas tends to
reflect game theoretical constructs more than human reality. "Rationality"
in human behavior comes to be defined as the clear identification of utility
goods and their effective pursuit.[9] This is even extended to the self-interested selection of a spouse.[10] Covenantal love, lasting commitments, and
abiding beneficence are stripped from the human scenario as game theo-
rists present it.

Early-school evolutionary biology does not allow for altruism between
nonrelatives, where "Tit for Tat" strictly applies. Any discernible love
toward nonkin who fail to reciprocate would be described as the toleration
of defection for some self-interested reason, such as the benefits of belonging to a supportive religious community or the hope of heavenly rewards.
There must be some "payoff equilibrium" in deviating from the system of
reciprocity. In essence, "Tit for Tat" gives the following outcome: "I help
you, but I expect that you will help me when we meet again. If you cooperate, I will; if you defect, I will." The rule for behavior is "On the first
move, cooperate, and then do unto others as they have done unto you." We
are told that this is how we do live and must live. Were we to follow the
rule, of course, we would live in a downward spiral of negative nit-picking.

Martin Nowak of Harvard University has criticized "Tit for Tat" as too unforgiving and ungenerous, resulting in a system where inevitable errors destroy cooperation.[11] He has developed a mathematical game model called "Generous Tit for Tat," which follows this rule: "On the first move, cooperate. If you cooperate, I will continue to do so to. If you defect, then I will still cooperate a third of the time." In other words, "Never forget a bad move, but be somewhat forgiving." This model represents a modification of direct reciprocity that, while still based on strictly self-interested assumptions of agency, nevertheless moves closer on the external level to a generous exchange. There is, ultimately, nothing unselfish about "Generous Tit for Tat," but it is suggestive in offering a context in which genuine altruism might more readily spill over into the nonkin group. Moreover, it shows that selfishness can get us to the point of a limited generosity of a sort, although still in the form of self-interested beneficence.

Both "Tit for Tat" and "Generous Tit for Tat" are based on direct reciprocity, but both include indirect reciprocity, whereby helping the other in the absence of an obvious "payoff" does result in an enhanced reputation. The ostensible altruist is really purchasing a reputation, from which he or she stands to gain in the long term. It is the case that we devote considerable time and energy to inquiry about the reputation of others. If rationality is ultimately defined by these game theories, then those who engage in random acts of kindness, unless seeking reputations by virtue of some calculated modest degree of generosity, have taken leave of their senses or relegated their unavoidable selfishness to the ethereal realm of immortal "payoff." In the final analysis, neither "Tit for Tat" nor "Generous Tit for Tat" can explain the deep morality of self-sacrificial giving to others, which as a human reality can be taken at face value.

THE PRESUMPTUOUSNESS OF EARLY-SCHOOL EVOLUTIONARY BIOLOGY

Returning now from game theory to further reflection on early-school evolutionary biology, Denis Alexander states that all reciprocal altruism is a "mechanism whereby the reproductive success of an individual is promoted by helping others."[12] One throws one's helpful behavior into a general pool expecting to draw from it as needed for the sake of self and kin. E. O. Wilson thus rightly designates reciprocal altruism as *soft core*, because it is "ultimately selfish."[13] He extols this form of altruism, which is actually not altruism at all. One must seriously wonder why the expression "recip-rocal altruism" is used, since self-interested cooperation is not resonant with altruism in the normal sense of the word. Wilson much prefers soft-core altruism, which is only enlightened egoism; *hard-core* or genuine altru-ism, which is kinship-based, strikes him as being volatile and irrational in any wider context. His preference is for a Rawlsian social contract.

In a rigidly contractarian ethical interpretation of early-school evolu-tionary biology, Richard D. Alexander, an early-school thinker, writes thus:

> Ethics, morality, human conduct, and the human psyche are to be understood only if societies are seen as collections of individuals seeking their own self-interests (albeit through use of the group or group cooperativeness, and given that, in historical terms, the individual's self-interests can only be realized through reproduc-tion, by creating descendents and assisting other relatives). In some respects these ideas run contrary to what people have believed and been taught about morality and human values: I suspect that nearly all humans believe it is a normal part of the functioning of every human individual now and then to assist someone else in the real-ization of that person's own interests to the actual net expense of those of the altruist. What this "greatest intellectual revolution of the century" tells us is that, despite our intuitions, there is not a shred of evidence to support this view of beneficence, and a great deal of convincing theory suggests that any such view will eventu-ally be judged false.[14]

According to this view, if even a modicum of indiscriminate or uncalcu-lating beneficence exists in human society, it is nevertheless grounded in self-interest and reputational investment.

What is it that early-school evolutionary psychology wishes to tell us about human nature? David P. Barash, a professor of psychology and zoology at the University of Washington, is especially vivid in his portrayal of human nature. Barash begins with the full spectrum of animal evolu-tion, in which "fitness is a manner of numbers, of cold-heartedly assessing genetic profit margins, in which the miserly, evolutionary Scrooge only wants to know how many genetic descendants are catapulted into the future."[15] In short, every organism pursues its genetic self-interest and en-gages in altruistic behavior only insofar as it is consistent with said interest.

Therefore, "what appears to be altruism at the level of bodies can be plain old selfishness at the level of genes."[16] For altruism is never random or indiscriminate, but a matter ultimately limited by *inclusive fitness* or reproductive success—not only of the individual, but of those kin who carry all or some of its genes. Thus, we are placed in the iron grip of a "predictable bias toward kin: relatives over nonrelatives, and closer relatives over more distant ones."[17] When a zebra is attacked by a lion, other zebras come to the rescue because zebras live in extended families and are therefore genetically related; wildebeests do not exhibit such rescuing behavior because they travel in a horde of nonrelatives. Genetic relatedness is the key to altruism, thereby unmasking all such altruism as genetic selfishness.

And does altruism exist in humans outside of kinship? Barash indicates that it really is not altruism at all, but a facade intended to manipulate one's audience for future gain. Any genuine group-level altruism is due to a higher-than-average degree of genetic relatedness within the group. Thus, "the secret wish of the Good Samaritan is that his or her altruistic act would be discovered and publicly acknowledged, but without the self-advertisement that might suggest the altruism is only being done for show or to attract mates, and thus may not be genuine." In all altruism, "bodies behave so that genes can get their way." Even the Hebrew scriptures' injunction to "love your neighbor as yourself" (Leviticus 19:18) "almost certainly means literally *neighbor*, that is, someone who is nearby, not just anybody and everybody, but an in-group member. Loving one's neighbor may turn out to be less a matter of true altruism than of genetic selfishness."[18]

What of cases where we behave altruistically toward nonkin? We are only designed to be altruistic when there is "a good chance that sometime in the future, the tables will be turned—literally!"[19] Altruism toward nonkin must always pay off genetically in the long run, and "hell hath no fury like an 'altruist' (actually, would-be reciprocator) whose generosity is not repaid."[20] Our large brain size may well be the result of the demands of score keeping and the identification of cheaters in the game of reciprocity. We are "natural-born reciprocators," and "we all enter into reciprocal relationships, pretty much every time we interact with other people." Thus, "social contracts of one sort or another operate in our most intimate, day-to-day lives, and when all parties abide by the expectation of reciprocity, everyone is far more likely to be satisfied than if someone assumes that he or she is entitled to 'get' without 'giving.'"[21] Human social behavior boils down to kin selection and reciprocity: "With whom, for example, do people interact? Offspring, mates, friends, and those with whom we do business or exchange information."[22] The reciprocal contract, which explains all social interaction outside of genetic relatedness, follows the pattern of game theory of being altruistic at the start, and then following the lead of the other. If he or she is altruistic in return, splendid; otherwise, don't be a "sucker." Barash believes this is so integral to human nature that those who seem to break it through random acts of altruism without apparent interest in reciprocity are, in fact, really interested in enhancing their reputations for eventual genetic gain. There will be an "indirect" return from a grateful society, a "payoff in being *perceived* as an altruist."[23] Even the holy men of Buddhism and Hinduism, seen as the epitome of altruism, are making a mere "public facade" with their begging bowls. These "cheaters and deceivers" are looking for a free meal.[24] Barash's statement here is, of course, simply absurd, for it has absolutely no foundation in any careful study of these holy men.

Matt Ridley, another accurate student of early-school thought, writes, "Babies take their mother's beneficence for granted and do not have to buy it with acts of kindness. Brothers and sisters do not feel the need to reciprocate every kind act. But unrelated individuals are acutely aware of social debts."[25] And Ridley celebrates the imperative of reciprocity, which he sees as part of our natures and as the building block of social experience and neurological capacity: "To play the reciprocity game, they need to recognize each other, remember who repaid a favor and who did not and bear the debt grudge accordingly."[26] Reputation counts, good reciprocators finish first, and defectors are eventually punished even if single defections can be forgiven. "Think about it," writes Ridley, "reciprocity hangs like a sword of Damocles over every human head."[27] Emotions such as gratitude, guilt, indignation, compassion, and the like are the tools of a reciprocal creature. Moral emotions, which are universal, exist "to enable us to pick the right partner to play the game with."[28] And while we admire and are fascinated by pure altruists, we "simply do not practice what we preach. This is perfectly rational, of course. The more other people practice altruism, the better off for us, but the more we and our kin pursue self-interest, the better for us. This is the prisoner's dilemma. Also, the more we posture in favor of altruism the better for us."[29] Human beings "are an extremely groupish species, but not a group-selected one. We are designed not to sacrifice ourselves for the group, but to exploit the group for ourselves."[30] While religions teach universal beneficence, in reality they live by "us versus them," for religions "have thrived to the extent that they stressed the community of the converted and the evil of the heathen."[31]

In summary, early-school evolutionary biology instructs us, in the words of Melvin Konner, that there are "biological constraints on the human spirit."[32] The iron laws of evolutionary biology are not, however, ones that I think we humans live by. We often relate to other human beings outside the law of reciprocity by reaching out, in a sense of shared humanity, to anyone and everyone who is in genuine need. As Batson has shown, we engage in "random acts of kindness" along the compassion-altruism axis.

Prince Peter Alexeyevich Kropotkin, whose classic work on evolution, *Mutual Aid: A Factor in Evolution,* is often cited by contemporary scholars, was careful not to impose any iron laws on human nature. Kropotkin was open-minded about the possibility of expanding genuinely benevolent social impulses beyond in-group boundaries. Moreover, he saw religion and spirituality as being important to this expansion. He recognized that culture can affect biological instinct, principally through the teachings of religions, including Christianity and Buddhism:

> The higher conception of "no revenge for wrongs," and of freely giving more than one expects to receive from his neighbors, is proclaimed as being the real principle of morality—a principle superior to mere equivalence, equity, or justice, and more conducive to happiness. And man is appealed to be guided in his acts, not merely by love, which is always personal, or at the best tribal, but by the perception of his oneness with each human being.

While mutual aid and reciprocal support are the "undoubted origin of our ethical conceptions," ethical progress occurs through its universal extension and release from constraints.[33]

An Alternative Image of Human Nature: New-School Evolutionary Biology

There is deep division within evolutionary biology as the early-school dismissal of group selection has been seriously questioned. The new-school evolutionary biology allows for group selection. Group selectionism holds that the group can come first, making genuine altruism more generally available across society. In the words of the two most important defenders of contemporary group selection theory, Elliott Sober and David Sloane Wilson, "Genuinely altruistic traits that evolve by group selection became an endangered species in evolutionary biology during the 1960s and 1970s, just as psychological altruism has long been an endangered species in the social sciences."[34] They argue that the rejection of psychological or motivational altruism in psychology has been reinforced by the view that evolutionary altruism (group selection) is absent from nature: "If evolutionary altruism is absent in nature, why should psychological altruism be present in human nature?"[35] They point out, however, that George Williams himself has moved away from a strict individual selection model. Moreover,

> The idea that human behavior is governed entirely by self-interest and that altruistic ultimate motives don't exist has never been supported by either a coherent theory or a crisp and decisive set of observations. The entire debate has been characterized by an intellectual pecking order in which an egoistic explanation for a given behavior, no matter how contrived, is favored over an altruistic explanation, even in the absence of empirical evidence that discriminates between the two approaches.[36]

In demonstrating the arguments for group selection in evolution, Sober and Wilson do not think that they have presented a theory that supports universal benevolence: "Group selection does provide a setting in which helping behavior directed at members of one's own group can evolve; however, it equally provides a context in which hurting individuals in other groups can be selectively advantageous." Nor do they wish to imply that "everyone has a thoroughgoing and saintly dedication to helping others— that people always treat the well-being of others as an end in itself and never think of their own welfare." Rather, they wish to show that "concern for others is *one* of the ultimate motives that people *sometimes* have."[37] This is a move in the right direction.

After an extensive critique of evolutionary theory and, separately considered, the social science of altruism and egoism, these authors conclude as follows: "We have shown that the wholesale rejection of group selection—and therefore of group-level functionalism—by evolutionary biologists during the 1960s was misconceived and has not withstood the test of time." Separately evaluated, "the case against evolutionary altruism has already crumbled," and the case against psychological altruism is weak and crumbling. Genuine altruism "can be removed from the endangered species list in both biology and the social sciences."[38]

If group selection theory is true, it suggests that the altruistic behavior seen in groups (from flocks to tribalism) is directed toward the group as a whole and its survival vis-à-vis competing outside groups. Sober and Wilson consider Charles Darwin to be one benchmark of support for this thesis.[39] There is no longer any reason to accept the early-school theory that genuine group altruism is illusory. We need not accept the basic parameters of evolutionary biology that state that social relations beyond kin are manifestly self-interested and can be nothing else. Our cooperation in groups need not be viewed as intractably egoistic.[40] (It is, however, likely that when natural selection acts at the group level, we see heightened tendencies toward aggression between groups.[41] Competition is transposed from individuals to groups.)

Group selection is an idea that Darwin himself thought quite plausible, as did other biologists such as V. C. Wynee-Edwards and Warder Clyde Allee. Despite the wide influence of Williams, Trivers, and Dawkins, group selectionism has reemerged—that is, the idea that in nature groups as a whole can become well adapted and that the forces of natural selection can rise above the individual to group success, allowing genuine altruism to evolve.

What does one conclude about human altruism toward nonkin? Robert H. Frank summarizes a good deal of research on human helping behavior toward strangers in New York City, a context more or less devoid of "Tit for Tat" and other forms of reciprocal altruism in interactions among strangers: "The obvious difficulty for the self-interest model is that, in virtually every study, New Yorkers do not behave in the predicted manner." People behave constantly in ways that transcend "Tit for Tat," which is "not genuinely *altruistic* behavior. Rather it is, like reciprocal altruism, a straightforward illustration of *prudent* behavior—enlightened prudence, to be sure, but self-interested behavior all the same."[42] Nor does Frank find any evidence to suggest that parental behavior is not often deeply reflective and genuinely altruistic as it cares for a future generation. Similarly, Paul R. Ehrlich marshals considerable experimental evidence to conclude that "empathy and altruism often exist where the chances for any return to the altruist are nil."[43]

Group selection and group altruism remain controversial, and do suggest strong in-group loyalties and out-group animosities. Yet there is much to be said for this theory as a serious rival to the too-easy assumptions made by the early-school evolutionary biologists as to the ubiquity of self-interest in all nonkin human relations.

CONCLUSIONS

Many scientists assume that altruism outside of the kin context must be subjected to innumerable intellectual assaults that would never be applied to the reality of egoism. They reveal a distorting bias.

Science could begin with the following alternative statement: "Scratch an egoist and watch an altruist bleed." It seems to me that we often try to convince ourselves that human nature is worse than it is, and that our generous other-regarding tendencies are somehow false. We want to live in the light of "tough" self-centered and aggressive images of human fulfillment as these are conveyed through popular culture. It is deemed embarrassing by some to be labeled as a "do-gooder." So we try to inhibit our altruism toward all humanity. This chapter, as well as the one that preceded it, are intended to encourage a common sense confidence in our everyday motivations to tend to the neediest, to care for the stranger, and to do good to all humanity without exception. Have confidence that there is an angelic aspect in a divided human nature. Plausibly, this side of human nature can be lifted up by the surprising energy of unlimited love. If we are to have a human future at all, it can only emerge from a love for humanity that transcends self, kin, and group to embrace all. This expansion defines both the moral and the spiritual points of view as bequeathed to us from all those perennial minds deemed wise in one tradition or another. There is no other real option.

6

THE PARENTAL AXIS

AND THE ORIGINS OF LOVE

WHILE IN THE PREVIOUS chapter I have been somewhat critical of early-school evolutionary biology, I have not been dismissive. In fact, these thinkers have brilliantly uncovered the seed from which most abiding and genuine altruism toward all humanity must grow—perhaps first through genetically related groups, and then to larger groups, and finally, given the right cultural, intellectual, and spiritual influences, to all humanity. In other words, the most potent and universal form of altruism is on the parental axis, as any observer of human behavior across all cultures and civilizations must note. Yet human nature is malleable to the extent that this altruism can increase in extensivity.

As was argued previously, while Dawkins and others have tainted parental altruism with the color of "selfish genes" in the background, genes are absolutely incapable of being "selfish." Holmes Rolston explains, "The production and defense of natural kinds [species] are what is ultimately involved in the alleged 'selfishness' of these genes. The historical evolution and reenactment of individuals instantiating the diverse natural kinds cannot be evil." He further notes that the use of the metaphor of "selfishness" is "loaded with pejorative overtones" that are unreasonable.[1]

Even Dawkins qualifies his language by stating that selfish genes are perfectly capable of producing genuinely unselfish beings on the kin level. It is useful to clearly delineate a parallel universe of phenotypical and genuine altruism alongside the underlying genetic replicators. This is not to contest the importance of the genetic "replicators," but as Steven Pinker writes regarding parental love, "What is selfish is not the real motives of the person but the metaphorical motives of the genes that built the person." More pointedly, Pinker continues:

> The confusion comes from thinking of people's genes as their true self, and motives of their genes as their deepest, truest, unconscious motives. From there it's easy to draw the cynical and incorrect moral that all love is hypocritical. That confuses the real motives of the person with the metaphorical motives of the genes. Genes are not puppetmasters; they acted as the recipe for making the brain and body and then they got out of the way. They live in a parallel universe, scattered among bodies, with their own agendas.

Regrettably, adds Pinker, the selfish gene theory presented by Dawkins confuses these parallel universes, and seems to imply that human altruism is merely a facade. As Pinker rightly insists, "I think moralistic science is bad for morals and bad for science. Surely paving Yosemite is unwise, Gordon Gekko is bad, and Mother Teresa is good regardless of what came out in the latest biology journals."[2]

Love, compassion, and empathy are genuine and real, even if they emerged in order to promote the well-being of those endowed with our own genes. The key ethical question is how to extend love, compassion, and empathy to all humanity, as Mother Teresa and countless others seem to have done in ways that transcended any and all genetic restraints on the human spirit. There does come a point where efforts to explain away such extensive love, rather than accept it at face value, approach absurdity.

THE EVOLUTIONARY CENTRALITY OF PARENTAL LOVE

Parental altruism is a partial, indirect, but significant reflection of the very nature of unlimited love. Human parental altruism has its roots in a remarkable evolutionary process that seems to correlate with the phenomenon of prolonged infant dependency requiring parental care. In mammals especially, there is a correspondence between lengthened infancy and parental altruism, especially in its maternal form. Such altruism preserves offspring and must, therefore, be heavily favored by natural selection; there is a movement toward cherishing the life of another more than one's own. The word "mammal" comes from the Latin word *mamma*, which refers to the breast; mammals as a class are characterized by female secretion of milk to feed the young. In human beings, the infant's protracted dependency, dawning consciousness, and cuteness open up a narrow evolutionary horizon for tender altruistic sentiments and unselfish ends. In the midst of a clashing struggle for life, the undertone of altruistic affection and the possibility of moral nobility emerges.

It is refreshing to take leave of contemporary theorizing in evolutionary biology and return to Darwin, who saw tremendous moral importance in maternal parental sentiment. In his *The Descent of Man*, written in the 1870s late in his career to address human evolution and nature, Darwin pays eloquent homage to the human capacity for altruistic emotions: "With respect to the origin of the parental and filial affections, which apparently lie at the base of the social instincts, we know not the steps by which they have been gained; but we may infer that it has been to a large extent through natural selection."[3] He understood the insular tendencies of such affection, but he also refused to underestimate its plasticity with regard to nonkin:

A young and timid mother urged by the maternal instinct will, without a moment's hesitation, run the greatest danger for her own infant, but not for a mere fellow-creature. Nevertheless many a civilized man, or even boy, who never before risked his life for another, but full of courage and sympathy, has disregarded the instinct of self-preservation, and plunged at once into a torrent to save a drowning man, though a stranger. . . . Such actions as the above appear to be the simple result of the greater strength of the social or maternal instincts than that of any other instinct or motive; for they are performed too instantaneously for reflection, or for pleasure or pain to be felt at the time.[4]

Darwin further emphasizes the expansion of familial and motivationally altruistic sentiment beyond its insular embryonic context: "Nevertheless, besides the family affections, kindness is common, especially during sickness, between the members of the same tribe, and is sometimes extended beyond these limits."[5] Building upon parental altruistic sentiments with Kantian confidence in the power of human reason to appreciate the importance of all humanity, Darwin writes:

As man advances in civilization, and small tribes are united into larger communities, the simplest reason would tell each individual that he ought to extend his social instincts and sympathies to all

members of the same nation, though personally unknown to him.
This point once reached, there is only an artificial barrier to prevent
his sympathies extending to the men of all nations and races.[6]

Darwin appears to move from 1) a deep appreciation of parental altruistic sentiment to 2) a clear extension of such sentiment to group altruism
to 3) a Kantian expansion of such sentiment to all humanity. While Darwin
refers specifically to Kant and the emergence of human freedom through
the power of reason, he opposed Kant's denigration of the importance of
human instinctual inclination and emotion. He appears to rest the "moral
sense" in a balanced dyad of emotional inclination and reason. Furthermore, the expansion of altruistic sentiment is grounded in group selection theory: "A tribe including many members who, from possessing in a
high degree the spirit of patriotism, fidelity, obedience, courage, and sympathy, were always ready to aid one another, and to sacrifice themselves
for the common good, would be victorious over most other tribes; and
this would be natural selection."[7] A year after publishing the first version of
The Descent of Man, Darwin presented another essential study entitled *The
Expression of the Emotions in Man and in Animals*. Here he further defined the
centrality of altruistic emotion under the rubric of "Love, tender feelings,
etc.," with immediate focus on maternal sentiment.[8]

William McDougall, a founding father of social psychology, borrows
Darwin's language in his classic 1908 text on the subject. He distinguishes
Darwin's "tender emotion" from the more limited and egoistic processes
of sympathy. Sympathy, as he understands it, is a form of pain (or pleasure) invoked by the spectacle of the other that results in our wanting to
"turn our eyes and thought away from the suffering creature." In an
intriguing reference to the parable of the Good Samaritan, he writes:

> No doubt the spectacle of the poor man who fell among thieves
> was just as distressing to the priest and the Levite, who passed by
> on the other side, as to the good Samaritan who tenderly cared for
> him. They may well have been exquisitely sensitive souls, who
> would have fainted away if they had been compelled to gaze upon
> his wounds. The great difference between them and the Samaritan
> was that in him the tender emotion and its impulse were evoked,
> and that this impulse overcame, or prevented, the aversion natu-
> rally induced by the painful and, perhaps, disgusting spectacle.

McDougall regularly refers to the "tender emotion" that "draws us near to the suffering and the sad," allowing for "disinterested beneficence," and associates it most nearly with maternal tendency: "The response is as direct and instantaneous as the mother's emotion at the cry of her child or her impulse to fly to its defense; and it is essentially the same process."[9]

It is clarifying to reflect now more fully on a distinction made in 1978 by E. O. Wilson in his thoughts on human nature, predating the publication of Dawkins's *The Selfish Gene*:

> To understand this strange selectivity and resolve the puzzle of human altruism we must distinguish two basic forms of coopera- tive behavior. The altruistic impulse can be irrational and unilater- ally directed at others; the bestower expresses no desire for equal return and performs no unconscious actions leading to the same end. I have called this form of behavior "hard-core" altruism, a set of responses relatively unaffected by social reward or punish- ment beyond childhood. Where such behavior exists, it is likely to have evolved through kin selection or natural selection operating on entire, competing family or tribal units. We would expect hard- core altruism to serve the altruist's closest relatives and to decline steeply in frequency and intensity as relationship becomes more distant. "Soft-core" altruism, in contrast, is ultimately selfish. The "altruist" expects reciprocation from society for himself or his closest relatives. His good behavior is calculating, often in a wholly conscious way, and his maneuvers are orchestrated by the excruci- atingly intricate sanctions and demands of society.

In honeybees, kin selection is ubiquitous and therefore all altruism is hard core. For human beings, Wilson continues, things are different:

> But in human beings soft-core altruism has been carried to elabo- rate extremes. Reciprocation among distantly related or unrelated individuals is the key to human society. The perfection of the social contract has broken the ancient vertebrate constraints imposed by rigid kin selection. Through convention of reciprocation, com- bined with a flexible, endlessly productive language and a genius for verbal classification, human beings fashion long-remembered agreements upon which cultures and civilization can be built.[10]

Wilson clearly does find a hard-core or genuinely motivated altruism in human nature. In my view, this is the base metal of all human altruism, and the root of our capacity to surpass the self-interested reciprocal and cooperative limit. But remarkably, Wilson only wishes to denounce hard-core altruism based on kin selection as "the enemy of civilization," and the source of insularity and conflict along the lines of the Hatfields and McCoys. His hope for the human future lies strictly with soft-core altruism—that is, self-interested reciprocity: "Human beings appear to be sufficiently selfish and calculating to be capable of indefinitely greater harmony and social homeostasis."[11] Wilson has here set aside the potential value of "hard-core altruism," in part because he does not see its potential expansion through the group selection process that he has summarily dismissed.

It is not my point here to refute Wilson so much as to make clear the useful distinction that he makes between hard-core and soft-core altruism. What he refers to as soft-core altruism is, in the classical context of the distinction between motivational altruism and egoism, actually a form of egoism. (Such pseudo-altruism should be called by its real name.) The key point, for my purposes, is that for Wilson, genuine motivational altruism or hard-core altruism emerges from the kin context. If, then, we are concerned with establishing the human possibility of psychological altruism in humans, we must be particularly concerned with its roots in kinship (especially on the parent-child axis) and the malleability of this motivational capacity to wider social circumstances of nonkin, especially through the influence of culture, religion, and spirituality.

When the new-school evolutionary thinkers Sober and Wilson wished to defend the evolution of psychological altruism, they focused firmly on the parent-child axis: "We conjecture that human parents typically *want* their children to do well—to live rather than die, to be healthy rather than sick, and so on."[12] They do not ignore the fact that parents may have other competing desires that influence how they treat their children, and that may even result in infanticide under specific conditions. Sober and Wilson select parental care as so central in understanding the roots of human compassion because it is undeniably shaped by natural selection, and because they believe that its "motivational basis generalizes to helping directed to individuals other than one's offspring."[13] They surmise as follows:

> Indeed, helping one's children and helping individuals other than one's offspring are probably not totally separate—like body weight

and eye color—that evolved independently of each other. The quality of the mother-child bond appears to be a crucial predictor of the empathy and prosocial behavior that the child exhibits later in life. This suggests that when selection favored parents who took care of their children, it thereby favored children who provided help to others.[14]

I will not review the arguments these authors present, except to indicate that their conclusion is that parental care "is driven in part by altruistic motives,"[15] and that these motives are indirectly linked with broader social expressions of altruistic care.

Human beings are, technically, a "K-selected" rather than an "R-selected" species. R-selected species produce large quantities of offspring without parental care and investment; reproductive success involves producing sufficient numbers of untended offspring so that some will manage to survive the perils of nature. In contrast, K-selected organisms produce relatively few offspring but provide care, usually in the form of protection. The emphasis is on quality, not quantity. In evolutionary language, *Homo sapiens* is especially high in what is termed "parental investment."[16] As Robert Wright argues, our species is particularly high in "male parental investment," or "MPI":

> Fathers everywhere feel love for their children, and that's a lot more than you can say for chimp fathers and bonobo fathers, who don't seem to have much of a clue as to which youngsters are theirs. This love leads fathers to help feed and defend their children, and teach them useful things.[17]

A wide body of literature concludes that human males have high MPI because of a combination of the fact that 1) women have relatively narrow birth canals as a result of the narrow pelvis associated with walking upright; 2) the heads of babies became larger and larger to provide room for the evolving brain; and 3) as a result of 1) and 2), human infants are born very prematurely, requiring intense maternal investment (of the sort that hampers her food gathering). Men with high MPI were necessary for protection and provision. Maternal and paternal love are simply the emotional manifestations of parental investment. Natural selection invented parental love to confer benefits on offspring; while this is clearly a necessary and salutary development, it can easily result in the unfortunate overindulgence of children and disinterest in the fate of those outside the kin circle. Parental investment is not just the provision of external necessities, but the provision of love as an emotional necessity. No child who is unloved will thrive, and no child who does not experience being loved will be able to pass it on to others.

Moral and spiritual teachings may extol parental love as a natural and socially salutary aspect of procreation, but such teachings will also caution against moral myopia.

Altruism on the parent-child axis, then, while the inevitable point of origin of other-regarding love, emerged well targeted and intense. Love did not evolve in some weak and diffuse form. Universal love can only arise from the intense parental point of origin. Love widening in its narrowly targeted form to love of all humanity is the transposition of limited love into unlimited love.

There are certain deficits in parental love. Parental love, in addition to being potentially blinding and even obsessive, can manifest gender inequality. Parents might see better reasons to invest themselves in male rather than female offspring, and there is a history of almost exclusively female infanticide.[18] Furthermore, the very language of evolutionary biology in describing parental "investment" suggests a certain proprietary and manipulative tendency underlying parental love—a tendency that cannot be denied or ignored. Thus, religious rights of passage welcome the newborn into the community and entrust the wider set of believers with responsibility for his or her future, thereby protecting the child from the potentially dark side of parental love through the involvement of a wider group. Within the Christian tradition, baptism is the core right of initiation, and there was a time when "Godparents" had a very significant corrective role in a child's life. While evolved parental love is not without certain ambiguities, limitations, and impurities, it is nevertheless the closest love we have on earth to the disinterested altruistic love that many have associated with God through the use of parental metaphors.

If group-level selection is true, then we should find something like the direct and immediate empathic responsiveness of parent to child spilling over into wider domains of human social experience, especially in the emergence of large groups from genetically proximate origins. We should see the "tender emotion" enlarged to generalized other-regarding behavior.

Do we? As indicated earlier in this chapter, especially drawing on the writings of Batson, it appears that we do.

The rise of a general altruistic behavior from parental instinct suggests a highly selective dynamic energy modulating into a more extensive domain.[19] It also suggests that the presence of such a love instinct is a basic need in human development, and that this need pertains to human beings in general. Parents provide children with warmth, nourishment, and affection, and parental love is the most sustained motive in human life. Indeed, our highest knowledge of parental love is enlarged to cosmic dimensions in images of a parentally loving God, and of human beings as part of one family under God.

A "Phase Change" in the Direction of Unlimited Love

Parental love is generally narrowly focused and subject to certain forms of calculating self-interest and proprietary attitude. I do not wish to make it into something purer or better than it is. And yet it is the foundational evolved building block that, when uncoupled from our ancestral environment, can be refashioned into something that goes beyond its beginnings and becomes even more extraordinary than it is by nature.

Hans Jonas, in his monumental work entitled *The Imperative of Responsibility*, traces all human morality back to the human parent-child relation, to the natural feeling of responsibility for the utterly vulnerable but entirely valuable human infant brimming with promise. The "ought" manifest in the infant is the beginning of all moral sentiment and experience. Jonas interprets this parent-infant care as not only "the archetype of all responsibility," but "also its initial germ in the generic human condition."[20]

The historian John Boswell, in his history of adoption in the western world, writes of the centrality of parental love: "Everywhere in Western culture, from religious literature to secular poetry, parental love is invoked as the ultimate standard of selfless and untiring devotion, central metaphors of theology and ethics presuppose this love as a universal point of reference, and language must devise special terms to characterize persons wanting in this 'natural' affection."[21] The fact that this historian of western thought found the centrality of the metaphor of parental love compelling is testimony to the evolutionary arguments I have put forward here.

The idea of a "phase change" in human nature allowing for a whole new level of human relationality to emerge from building blocks that evolved on a lower stratum is not at all unscientific. Take the analogy, for example, of the emergence of the human mind from various preliminary elements of the brain. An explosion took place in human neurological organization between thirty and sixty thousand years ago in our species. *Homo sapiens* was first seen one hundred thousand years ago sharing the scene with both the Neanderthals and archaic *Homo sapiens*. Steven Mithen describes the evolution of language, technical intelligence, social intelligence, and natural history intelligence as a "whole series of cultural sparks that occur at slightly different times in different parts of the world between 60,000 and 30,000 years ago."[22] Suddenly, for example, we see the production of remarkably complex and beautiful ivory statuettes and cave paintings. The impressive horses, engraved owl, and rhinoceri paintings on the walls of the Grotte Chauvete (30,000 B.C.E.), for example, are splendid. Mithen comments that "although this is the very first art known to humankind, there is nothing primitive about it."[23] He sees nothing gradual about this development, as he compares the beauty of cave art to the great works of the Renaissance. Mithen attributes this human capacity to "new connections between the domains of technical, social and natural history intelligence."[24] Previously compartmentalized capacities in the brain could now interact, making possible a new magnitude of integrative creativity. At this same time, we see the emergence of religion, as evidenced by rituals that were intended to harness higher powers to bring about change in the natural world. Ritual activities, including religious burials concerned with afterlife, are evident in the anthropological record.[25] Mithen refers to a "big bang" in the history of human religious, artistic, and technical capacities due to a new fluidity of consciousness.

This explosion in human capacities is the dawning of a new morning in the evolution of life on earth. It is the dewy hour of sunrise in the springtime of consciousness. For the theist who, against the background of the anthropic principle, surmises some divine action mysteriously underlying the emergence of a creature who can now feel the presence of a Creative Presence, this is all a story of Cosmic Process. The roots of this new morning are found in the deepest foundations of the universe, in the far-off beginnings of the cosmos. The universe was "set up" to give rise to a creature whose evolution would now be in increments of neurological capacity and eventually of spirituality and culture. The highest goal of the Creative Presence was to establish background conditions in the universe that would eventually allow for the neurological capacities of the one creature with whose life that Presence might have relationship. Religious sentiment must be added to the differentiation of the human creature through the emergence of language, art, and technique. This sentiment would go on to shape history, art, and literature. Along with the first stammerings of speech and the dawning ethical discrimination between right and wrong came the earliest dim recognition of a presence in the universe greater than our own, in harmony with which rests our fuller well-being.

Is it not possible that, due to the complexity of the human brain and our capacities for self-transcendence and symbolism, the extraordinary love for all humanity could become a reality that unfolds from narrower evolved loves as these are taken over by the logic of equality and the spirit of unlimited love into a domain of infinite possibilities?

One of the leading philosophers of evolutionary theory, Elliott Sober, writes that "the ability to feel extended compassion is correlated with the ability to feel compassion toward close relatives, including one's offspring. . . . The present conjecture is that the psychological capacities that underwrite this advantageous trait have side effects. Individuals well attuned to the suffering of those near and dear have circles of compassion that potentially extend quite far afield." In other words, "A spin-off consequence of this evolutionary event is that human beings are inclined, in suitable circumstances, to feel compassion toward nonrelatives."[26] I think that Sober is correct in identifying the emergence of more generalized compassion with the parental and relational affect.

CONCLUSIONS

The world is too full of exemplars of unlimited love for us to think that love is heavily constrained or chained to the parental axis from which it emerged. Too many human beings have freed themselves to love all humanity with depth and commitment for us to accept the pessimistic argument that we cannot go beyond this narrow axis, or beyond intergroup conflict for that matter. Someday, the circle of unity under unlimited love will be unbroken. The latent potentialities for unlimited love in human nature are both unexplored and underappreciated. Neither mathematical models of "Tit for Tat" game theory, nor theories about how the purported "selfishness" of genes is somehow controlling of the human psyche, nor "group selectionist" ideas about how for every act of in-group kindness there must be a corresponding act of out-group cruelty are able to stand up before the reality of the extraordinary in human behavior that points toward unlimited hope for the future.

7

A Theological Interlude

on Parental and Unlimited Love

I N ITS FREEDOM from all the calculating relationships of everyday life, unlimited love seeks only the good of the beloved. While cooperative relationships of a reciprocal kind are extremely valuable and indeed necessitated by the realities of everyday organizational life, such relationships are made from a different base metal than unlimited love. Cooperation is important to a "successful" social life, and that reciprocation is often expected and deemed imperative. Yet there is another way of being in the world, without which our lives would be spiritually and morally impoverished. In the absence of unlimited love it is likely that all limited love would gradually expire and we would migrate into the domain of fear and retribution.

Despite scientific discoveries about the genetic malleability of human nature and the prospects for species enhancement through biotechnologies, the most promising avenue for improvement of the human condition lies in a deeper appreciation for and understanding of our capacity for other-regarding love. How can we gain more insight into the creative potential for love that manifests itself in individuals who spend their lives generously caring for others? Celebrated individuals, such as Gandhi or Mother Teresa, seem to realize this potential in extraordinary ways. And there are so many unrecognized persons in families and neighborhoods who regularly do good for others in humility and anonymity. How can the total energy of love in society and the world be increased? The discovery and development of love is the true hope for humanity, and promises infinitely more good than biotechnical enhancements.

In the previous chapter, I suggested that the least controversial and most probable foundation within human nature for unlimited love is the parent-child axis, which provides the evolutionary context for the gradual emergence of the capacities for compassion and love; these capacities can in turn evolve toward unlimited love through the influence of reason, culture, and spirituality. When we reflect back on the earliest part of our lives, we cannot recall the days of our infancy when we were entirely dependent on maternal love for milk, warmth, and bodily hygiene, and benefited greatly from the presence of a father whose love was salutary for both mother and child in countless ways. Mothers enjoy a place of honor in many cultures: in 1914, for example, President Woodrow Wilson declared the second Sunday in May to be Mother's Day. In some cultures mothers were even transformed into divinity. In Japan, the mother goddess was Izanami; in Egypt, Isis; in Finland, Mader-Akka; in the Congo, Mzambi.

THE FOUNDATIONS OF A NATURAL THEOLOGY

While the importance of parental love is not a new theological idea, it is always in need of articulation. John Fiske, the nineteenth-century philosopher-theologian of Harvard, wrote an important and classic essay in 1899 entitled *Through Nature to God*. In this work Fiske argued that the central fact in the evolution of *Homo sapiens* is prolonged infancy and its principle correlative, deepened parental love. Fiske, who was theologically reflective, stated that he found the then-new word "altruism" unattractive, but useful for rapport with the social sciences. He too saw the beginnings of altruism in maternal care as "no doubt the earliest; it was the derivative source from which all other kinds were by slow degrees developed." The instincts and emotions related to the preservation of offspring were favored and cultivated by natural selection as the alternative to extinction, achieving an apex in human parental love as "cherishing another life than one's own." Furthermore, "the capacity for unselfish devotion called forth in that relation could afterward by utilized in the conduct of individuals not thus related to one another."[1] The section in which Fiske develops this thesis is entitled "The Cosmic Roots of Love and Self-Sacrifice," for he believed that the emergence of such cherishing love resulted from the ethical trend of the universe and represented the pinnacle of cosmic creativity. With the lengthening of human infancy, more powerful affective capacities for compassion and love evolved. Fiske thus provided an interpretation of Darwinian evolution that was contrary to Thomas Huxley and Herbert Spencer, whose writings in the 1890s captured only the brutal aspects of the survival of the fittest.

Fiske's observations about human infancy and the evolution of compassionate love are no longer controversial. Parental care is extremely rare among invertebrates, where, as Susan Allport explains, "natural selection has favored the alternative strategy: production of large numbers of eggs." Mammals, however, are unique in that they incubate eggs internally in the uterus, give birth to live young, and have feeding built into their reproductive system through mammary glands. Thus, all mammals are "committed to a period of parental care after birth." The length of this period of care varies. The brain is fully developed at birth in precocial mammals, so the period of care is relatively short; it is relatively long in altricial species, where "brains of altricial young finish developing in the rich experiential context of the real world (rather than the dull, predictable context of the uterus or egg)." Allport notes that *Homo sapiens* is especially altricial:

> At birth, the brain of the chimpanzee, our nearest living relative, is one half of its full adult size, but the brain of a human infant is just one quarter. Three quarters of the human brain's development takes place in the real world, an adaptation that has given humans the ability to move into and inhabit every part of the globe and one that has demanded extraordinary amounts of parental care.[2]

This unprecedented infant and child dependence, the result of the combination of a large brain and a birth canal that can only tolerate a small head, meant that *Homo sapiens* would develop empathic, compassionate, and loving emotional qualities beyond those of any other species.

The need for compassionate love is deeply wired within us from birth and constitutes a fundamental requirement for full human flourishing. The capacity to give such love evolved within the limbic system of the brain, which is the seat of emotion, and was directed at the dependent infant. This emotional system exists because mammals split off from the reptilian line and developed a new neurological structure. As Thomas Lewis and his colleagues explain, "Detachment and disinterest mark the parental attitude of the typical reptile, while mammals can enter into subtle and elaborate interactions with their young."[3] The typical reptile lays its eggs and slithers away, although a few may be somewhat protective; mammals safeguard their young and care for them in ways that involve the emotional brain. Lewis and his colleagues summarize scientific research on the extraordinary importance of the human mother's empathy and love, and on the adverse consequences that result from its absence in this way: "An infant's brain is designed for ongoing attunement with the people predisposed to find him the most engaging of all subjects, the most breathtakingly potent axis around which their hearts revolve."[4] The love energy of parents must be considered an indispensable element in the life of any child and in the viability of any society (an obvious point understood long ago by Aristotle but not Plato). It is this energy, writ large, that lies at the natural core of unlimited love. Paradoxically, parental love is the most narrow manifestation of love, yet it is a form of love that is also general and ubiquitous in that all parents give it and all children receive it, unless precluded by powerful inhibitors. It is both intensely singular and remarkably universal.

Readers may question the idea of beginning a serious discussion of other-regarding love with comments on evolution and the brain, but where else can one reasonably begin? The three key parts of the brain evolved over millions of years in this sequence: 1) the brain stem or reptile brain; 2) the limbic system that covers the stem and in which compassionate love emerged on the mammalian maternal-infant axis; and 3) the neocortex, the seat of mentation. Emotions, which are clearly present in nonhuman animals, have a much longer history than previously thought, and are the seat of our capacities to love.[5] The limbic core of love remains the same whether it is directed toward one's child or more extensively. The emotion of other-regarding love surpasses anything that reason alone has to offer with regard to the care of others. The more we can learn about this emotion and its plasticity within its embodied neurological context, the more we will fulfill our highest potential.

I hypothesize, then, that the foundational emotional capacities for unlimited love were formed on the parent-child axis, awaiting only the influence of moral reasoning, culture, and spirituality to begin to expand them to include a wider domain. The human capacity for such love encompasses an "image of God," or at least an opportunity for God to mold that capacity into something better. There is no more pervasive and relatively constant expression of love than its parental form, which suggests that parental love hints at Unlimited Love. Is there a difference between human and transcendent love? Certainly in degree, but it is not clear that there is a difference in kind.

Yet what if it *is* a difference in kind, and therefore parental love is not a true base material to be elevated by divine Unlimited Love? Perhaps Unlimited Love must flow entirely from God's grace over and against every aspect of our human nature. Nonetheless, it makes sense to take the natural human empathic capacity as an important part of the mix. In this chapter I am less concerned with proving anything and more concerned with putting forward a plausible image of human nature in relation to divine love.

THE PARENTAL HEART IN THE IMAGE OF GOD

As a religious thinker, I see great value in the deepening of natural altruistic human tendencies and emotions that evolved largely on the parent-child axis and expanded outward in a manner that clearly jettisons concerns about mutuality and its various calculations. The most significant human enhancement is the deeper appreciation for and cultivation of the generous loving capacities that are already dormant within us. In his 1934 Rauschenbusch memorial lectures, given at the Colgate Rochester Divinity School, Reinhold Niebuhr noted two features of human nature that are invariably crucial to the moral life: "the natural endowments of sympathy, paternal and filial affection, gregarious impulses and the sense of organic cohesion which all human beings possess, and the faculties of reason which tend to extend the range of these impulses beyond the limits set by nature."[6] He believed that the rationalists of Stoic, Kantian, and utilitarian persuasion have not understood the significance of these endowments. "No rational moral idealism can create moral conduct," Niebuhr noted, because however much reason can provide criticism of insular tendencies and offer meaningful universal norms, it fails to provide "a dynamic for their realization." He concluded: "Thus the Stoics regarded the sentiment of pity as evil and in Kantian ethics only actions motivated by reverence for the moral law are good, a criterion which would put the tenderness of a mother for her child outside the pale of moral action."[7] The cleavage between the ideal of moral inclusivity for all humanity and the realities of tribalism and insular loyalties can only be overcome by the right kind of love:

Consequently, the Christian ideal of a loving will does not exclude the impulses and emotions in nature through which the self is organically related to other life. Jesus therefore relates the love of God to the natural love of parents for their children: "If ye then, being evil, know how to give good gifts unto your children, how much more will your Father which is in heaven give good things to them that ask him?" In its appreciation of every natural emotion of sympathy and human solidarity, the ethics of Jesus is distinguished from the ethics of rationalism.[8]

Clearly, there must be an important point of contact between unlimited love and natural altruistic loves.

While parental love undoubtedly has its evolutionary basis in the genetically driven procreative impulse and instinct, it is the most remarkable expression of abiding other-regarding emotion that we have available; as such, it could be viewed as nothing short of a miracle of nature and of nature's God. The relevant question is how it can be extended. This is not to say that unlimited love is simply parental love; it is parental love transposed in important ways to a new and distinctive key.

Unlimited love has its human roots in parental love, or what the Greeks termed *storge*. The most undisputed natural form of altruism is *storge*, and this is the closest thing to *agape* that human nature possesses. The love of a mother or father comes when the child feels deserted, lonely, desolate, and empty; it comes when the child needs to celebrate an accomplishment or express the joy of existence. Such love always comes in personal and intimate ways, for it knows the child by name. Whatever its ambiguities and impurities, it allows every child to feel beloved; it is perfectly particular to "this" very special child and yet perfectly ubiquitous in that all children are born under its canopy unless something has gone terribly wrong. Perhaps *agape* or unlimited love is God's *storge*, for like parental love, it even loves us when we are unlovable, when we have broken solemn promises, squandered opportunities, and broken cherished dreams. Unlimited love would, however, be a form of *storge* that is not inhibited by any of the inclinations that earthly parents have to limit their love.

Love, as the quintessence of the character of God, is regarded as axiomatic in monotheistic faiths. Jesus of Nazareth compares it to the love of parents for their children: "If you then, who are evil, know how to give good gifts to your children, how much more will your Father in heaven give good things to those who ask him?" (Matthew 7:11). He finds a chief symbol of divine love in the parental heart. Jesus often referred to the fatherly character of God, as in the Lord's Prayer (Luke 11:2–4), and to the motherly character of God, as in his lament outside of Jerusalem, in which he likens divine love to that of a hen gathering her brood (Luke 13:34, Matthew 23:37).

The appreciation of parental love as a signpost toward the divine is evident in Christian tradition as well. Julian, a fourteenth-century anchorite in the great medieval city of Norwich, England, and the mother of a large number of children before she turned to the religious life, recorded a series of "showings" from God. Because she sensed the female nature of Christ's suffering and the maternal aspects of divine love, she used many domestic images of motherhood in her descriptions. While these images were probably grounded to some degree in her own experiences before becoming an anchorite, one cannot disprove her claim that maternal love is objectively an aspect of divine love, nor reduce it to mere projection. She describes how divine love will not be discouraged or limited by human sin: "The mother may allow the child to fall sometimes and to be hurt for its own benefit, but her love does not allow the child ever to be in any real danger."[9]

One of the significant features of Celtic and Anglican Christianity is the recognition of understanding God through nature. C. S. Lewis appreciated *storge*, or "affection, especially of parents to offspring," as a heuristic key into the affectionate nature of divine love. He asserted that the fact that all animals seem to have this capacity does not lower its value.[10] Theologians have rightly contrasted the *agape* love of free self-emptying on behalf of the other with the reciprocal demands of *philia* and the appetitive nature of *eros*. With the exception of Lewis, however, they have not given *storge* any attention. Yet, continuity with *agape*, insofar as it exists, must lie in *storge*. As theologian Sally McFague has written, "the maternal metaphor is so powerful and so right for our time that we should use it." In an age of nuclear and ecological threat, she claims that a metaphor emphasizing nurture and "life as a gift" is necessary, for parental love "nurtures what it has brought into existence, and wants it to grow and be fulfilled." Evolutionary biology, as discussed in the previous chapter, would add weight to McFague's assertion that "parental love is the most powerful and intimate experience we have of giving love whose return is not calculated (though a return is appreciated): it is the gift of life as such to others."[11]

In emphasizing *storge* as theologically important, I have a highly idealized form of parental love in mind, stripped of any and all proprietary attitudes. There are those who dismiss parental love for its occasional distortions, but it is generally considered a supreme and indispensable human good; without it, we do not think that children can thrive, however resilient they may be. An analogy between the parental heart of God and the hearts of good human parents for their children can and should be asserted. This does not mean that images of parental love cannot be criticized insofar as the human phenomenon manifests flaws.

There is a spiritual tradition in Roman Catholicism that applies parental love to the world at large. For example, as Ivone Gebara notes, a rich tradition of "spiritual motherhood" exists among the Catholic "women religious." Gebara articulates a motherhood that leaps over institutional walls, one that can "beget though barren," characterized by "wise women whose close connection with life's realities enables them to listen to, feel for, advise and help those (of both sexes) who come to them." Gebara continues: "These women—widows, married, single parents, with few or numerous dependents—exercise a 'spiritual motherhood' among the people without giving their daily gift of life, their begetting of the Spirit, this or any other name."[12] This is not unlike McFague's suggestion that we "see ourselves as universal parents, as profoundly desiring not our own lives to go on forever but the lives of others to come into being."[13]

Parental love is not a self-indulgent mood of warm feelings; it is a sustained emotional tendency toward the good of another that is almost always replete with helping behaviors. The parent, either mother or father, actively responds to the needs of a child, whether expressed in the cries of an infant or the request for help with the down payment on a new house. Those who never become literal parents still possess all the affections of other-regarding love; they may, however, reasonably determine that emotional investment in offspring would be a distraction from higher callings of love for the very neediest.

ALL ANALOGIES LIMP

In drawing a loose analogy between divine *agape* love and the parental love of *storge*, I am aware that no analogy is perfect. Indeed, the discontinuities between *agape* or Unlimited Love and human loves of any sort are perhaps so considerable that no analogies should be attempted, as Karl Barth argued. In suggesting a point of continuity on the parent-child axis, I do not wish to be taken literally, because all theological language is only a human effort to approximate the mysterious nature of a presence in the universe that is higher than our own. I do not wish to make an idol of *storge*, for in associating divine love with parental love we run the risk of divinizing parenthood. Moreover, we must be cautious not to absolutize any of the ways in which we imagine God in the light of our limited human experience and in our own image.

But analogies are a crucial tool that enable us to develop our interpretation of the reality. Dorothy Emmet's definition serves well: "An analogy in its original root meaning is a pro-portion, and primarily a mathematical ratio, e.g., 2:4::4:x. In such a ratio, given knowledge of three terms, and the nature of the proportionate relation, the value of the fourth term can be determined. Thus analogy is the repetition of the same fundamental pattern in two different contexts." When we employ analogical reasoning we claim that a pattern that is obvious in one context applies in another, so analogy means "argument from parallel cases."[14] In analogical reasoning, we infer from the fact of two (or more) items sharing related and significant properties that they are more or less likely to share another relevant property that one of them is already known to have. Two entities are analogous if the relevant aspects of one entity are related in such a way that they agree with or correspond to the relevant aspects of the other. I have argued that we analogically attribute qualities of perfect parental heart to God because parental love is the most intense and abiding form of love with which we are familiar.

The conclusions of analogical reasoning lack the certainty of a geometric theorem. When we use the terms "analogy" or "analogical," we are referring to all those arguments that emphasize similarities between apparently dissimilar phenomena. It is important to keep in mind, for example, that divine unlimited love is ultimately a mystery that we can only speak of in epistemological humility, and that we will more easily approach a description of such love by saying what it is *not* rather than what it *is*. Yet our minds are highly analogical, as studies in artificial intelligence suggest. In analogical reasoning, one phenomenon is usually better understood than the other; what is better known is used to illuminate the less well known.

Theologically, the relationship between parent and child is known; from it, we illuminate the relationship of God to humanity.

Again, all analogies limp when it comes to the mystery of God, and there are those who would contend that divine unlimited love is utterly unlike any human emotion, even parental love. It is genuinely tempting to take this view, for human parental love has its inevitable imperfections and must be considered at best a vague and flawed reflection of Unlimited Love. Yet as Daniel Day Williams suggested, "We have seen in our study of the biblical view of love that if we say human experience throws no light on the meaning of the divine love, we are departing from the biblical mode of speaking about God."[15] While unlimited love as a divine energy cannot be fully grasped in terms of human experience, it is remarkable to see the lengths to which parents go in loving their children, even in the most challenging of circumstances.

Judeo-Christian tradition regards the idea of parental love as following a pattern of divine creation. The Jewish *hesed* or "steadfast love" is most frequently captured in the Hebrew Bible by the intimate familial love of parent for child. Paul Ramsey articulates the Christian continuation of such: "Of course, we cannot see into the mystery of how God's love created the world....Nevertheless, we procreate new beings like ourselves in the midst of our love for one another, and in this there is a trace of the original mystery by which God created the world."[16] Judeo-Christian thought elevates steadfast love that fuses parental creation and committed love to the sanctity of divine image.

Every natural theology presumes that something of the character of the Creator can be hinted at in creation, however partially. St. Paul wrote that God's attributes "have been visible, ever since the world began, to the eye of reason, in the things he has made" (Romans 1:20). According to Genesis, man and woman together are the image of God. Emil Brunner writes of the "divine pedagogy of creation." Building on the symbolic importance of threeness, Brunner sees in mother, father, and child "the trinity of being we call the human structure of existence."[17]

A natural theology of parental love is seldom made explicit. One articulation is found in John Burnaby's epilogue to *Amor Dei*. Burnaby describes the life of the child as beginning with complete dependence on parental love, from which he or she learns to be a *causa efficiens* of love. Burnaby continues:

If we now attempt to trace a corresponding pattern in the love which is God's own nature, it will be in the faith common to St. Augustine and St. Thomas, that His invisible things are understood through the things that are made, that there is a relation not of identity but of analogy, between the natural and the supernatural, between the changing and the changeless Good.[18]

It is God's parental love that is reflected in the human heart; it is the love of a parent that teaches a child what love is. Jesus of Nazareth appealed often to the parental metaphor in describing the nature of God and God's kingdom. Even sinful people give good gifts to their children (Matthew 7:11). The kingdom of heaven is like "a king who prepared a feast for his son's wedding" (Matthew 22:2). The parable of the prodigal son might well be called the parable of the merciful father. As Marie-Theres Wacker states, the eleventh chapter of the book of Hosea, often "celebrated by biblical scholars as the song of God the Father's love for Israel," in fact does not use the word "father." The parental activities in Hosea used to describe God are "a mother's everyday activities."[19]

Through the experience of parental love one begins to understand, however imperfectly, God's unlimited and personal love for each and every human being. If human parents tend compassionately to their children, then how much more must God? Indeed, for those who accept the parental analogy between human beings and God, a process theism in which God suffers in *pathos* from the waywardness of all human beings seems inevitable.[20] It is this sense of divine *pathos* that Abraham Heschel thought shaped the lives of the Hebrew prophets.[21] Love can be unlimited and abundant on one level and yet not entirely free of suffering.

Parental love in the human creature has a degree of inconsistency that surely falls short of unlimited divine love. Psychiatrist Willard Gaylin writes: "Love is not to be equated with the genetic bonding of animals—it is less and more. It is surely less immutable, as is evident from the number of neglecting parents and even more so from the number of barbaric parents who beat and batter their children."[22] Because parental love can be withheld or negated, there is good reason to emphasize the religious images and symbols that encourage parents to love their children.

The Roman Catholic tradition contains profound resources on parental love. In general, Catholic ethics has maintained the Augustinian-Thomistic readiness to interpret such love positively. In a remarkable but largely over-looked French study on love *(aimons nos freres),* Louis Colin refers to the child as "at the confluence of two hearts, of two beings who, mingling and merging their tenderness, pour it out afterwards on this little being." He adds: "Of all the human affections, those of fathers and mothers seem to be the most instinctive and incoercible. The very animals have a sort of uneasy, passionate solicitude for their progeny which sometimes moves us. The Carmelite of Lisieux felt her eyes moistening at the sight of the hen warming her chicks under her wings." Catholic natural law tradition understands, argues Colin, that "it is there in the mysterious depths where a human life is being shaped, in the very womb of the woman, that mater-nal love is first born." It views parental love as natural and at the same time as "a participation" in God's parental being.[23]

In the final analysis, parental love is reasonably close to being paradig-matic for *agape,* as long as it respects the freedom and maturity of the other. There is no human love that runs deeper or more powerfully. This love only awaits a breakthrough *(metanoiai)* to all humanity without exception. This is, of course, not natural in the simplistic sense of the ripple effect that flows outward from a pebble cast into a pond. There are those who see all natural loves as inherently selfish, and all genuine love for humanity as the result of an influential divine energy in the form of grace. I believe, however, that there are important continuities to be noted, and that grace works both with and against nature. The substrate of human nature is not entirely recalcitrant to Unlimited Love.

CONCLUSIONS

Unlimited love does seem to break into the lives of celebrated altruists, sometimes in an absolute reversal of direction through experiences of spir-itual crisis and sudden spiritual conversion. We are amazed by such divine grace. But even in such cases, the affective substrate must be present. Unlimited love is more often a matter of gradual, if difficult, growth over the journey of life, sometimes reaching its peak after insights into one's mortality and frailty. But however the narrative of love unfolds in any single life, unlimited love must reform and reshape an existing human substrate.

Among the positions of Protestant theologians, Paul Tillich's integra-tion of *agape* and less universal forms of love is extraordinarily insightful.

He rejected the either/or approach to *agape* and other loves, preferring integration instead. *Agape*, writes Tillich, "cuts" into friendship and elevates it from the "ambiguities" of self-centeredness.[24] He continues:

> Again, *agape* does not deny the preferential love of the *philia* quality, but it purifies it from a subpersonal bondage, and it elevates the preferential love into universal love. The preferences of friendship are not negated, but they do not exclude, in a kind of aristocratic self-separation, all the others. Not everybody is a friend, but everybody is affirmed as a person.[25]

Such friendship is not aristocratic or arrogant; rather, it exhibits the qualities of humility and openness. I would modify Tillich by replacing *philia* with *storge*, and by paraphrasing: "Again, *agape* does not deny the preferential love of the *storge* quality; it purifies it from a subpersonal bondage, and it elevates preferential love into universal love. The preferences of parental love are not negated, but they do not exclude all others. Not everybody is one's child, but everybody is affirmed as such."

In Henri J. M. Nouwen's reflections on the meanings in Rembrandt's painting "The Return of the Prodigal Son" (Luke 15:11–32), he focuses on the hands of the father. The father emanates forgiveness upon the prodigal's return, manifesting a love that is too great for force or constraint, a love that gives the son freedom to reject or return it. The father was powerless to prevent the son from running away to a distant country. Nouwen sees all the essentials of a constructive theology in this father and his relationship to his son:

> Here is the God I want to believe in, a Father who, from the beginning of creation, has stretched out his arms in merciful blessing, never forcing himself on anyone, but always waiting; never letting his arms drop down in despair, but always hoping that his children will return so that he can speak words of love to them and let his tired arms rest on their shoulders. His only desire is to bless.[26]

Nouwen's insights into the hands of the father are especially significant. Several art critics have commented that the left hand of the father is masculine and probably the artist's own, while the right hand is distinctively feminine. As Rembrandt captures him, the father is not only the great patriarch, but mother as well, touching the son with masculine and feminine hands. Observing the actions of both hands in the painting, Nouwen adds: "He holds, and she caresses. He confirms and she consoles. He is, indeed, God, in whom both manhood and womanhood, fatherhood and mother-hood, are fully present." There is unconditional love from a God who is "Father as well as Mother."[27] The return of the prodigal was cause for cel-ebration after the father's long suffering. In these images, Nouwen finds the analogical beginnings of a very simple but profound theology, that divine Unlimited Love reflects maternal and paternal aspects.

Filial love can never be demanded, only elicited. The parent is endowed with a love of children that makes it normally a pleasure to contribute to their welfare. But the child does not have a corresponding natural love for the parent. Filial love, in contrast to parental love, must be prompted and nurtured. Daniel Mark Epstein makes this astute observation in his study of the "natural history of the heart": "We are born from total darkness into a blinding light in which we cannot distinguish ourselves from mother or anything else in nature. Yet even before we recognize ourselves or our surroundings, love has been working on us for some time." Epstein reflects as far back as he can on his childhood and notes, "the first experience of love is being loved, by our parents." He further suggests that early mother love, in particular, elicits the initial return of love from the child. "As I got older," he concludes, "I became more conscious of my feelings, but the basic emotion of filial love did not change after childhood."[28]

I wish to conclude this discussion of parental love with a story that cap-tures the essence of this love transposed beyond the confines of the par-ent-child axis, in this case as manifested in the relationship of an old man to an adolescent. In one of his last novels, *The Adolescent (or A Raw Youth)*, published in 1874, Dostoyevsky focuses on the fact that parental virtue is deeply causative of filial response. Although the book was not well-received and still is not one of the author's best known, it is a classic study on death and growing old. In *The Adolescent,* Makar Evanovich Dolgoruky is a former serf, now old and gray and the legal husband of the mother of "the adolescent." "The adolescent," Arkady Dolgoruky, narrates his dia-logue with Makar, who offers this description of virtuous dying: "So a pious old man must be content at all times and must die in the full light of understanding, blissfully and gracefully, satisfied with the days that have been given him to live, yearning for his last hour, and rejoicing when he is gathered like a stalk of wheat unto the sheaf when he has fulfilled his mysterious destiny."[29] Arkady notes of Makar that "there was gaiety in his heart and that's why there was beauty in him. Gaiety was a favorite word of his and he often used it."[30] Makar seems to rejoice in the existence of things around him, whether human or nonhuman, animate or inanimate. His appreciation of the mystery of the world moves him far beyond a merely utilitarian relationship with the people and objects he encounters. He does not seek to turn the world to his advantage, and values things and people simply on the basis of their being. Arkady evidently feels this "gaiety" of heart, this mystical love for the world that is so different from mere aesthetics. Arkady continues: "Moreover, I'm sure I'm not just imagining things if I say that at certain moments he looked at me with a strange, even uncanny love, as his hand came to rest tenderly on top of mine or as he gently patted my shoulder."[31] Love is Makar's chief virtue.

The phenomenon of parental love as it emerged in this old man is profound and represents a high point with regard to the potential of human nature. This form of love is the point of connection between the vast domain of evolution and the cosmic presence of Unlimited Love. The prolongation of human infancy due to the enlargement of the human brain is perhaps the most widely accepted external explication of the presence in humans of a capacity for deep and enduring love; the internal reason is that God could work on the foundation of such love.

The whole travail of evolutionary nature is to bring forth the rough substrate for unlimited love in the human heart. The instincts that contributed to the preservation of offspring in various species reached a peak in human parental love, which is the ultimate natural form of cherishing another life. It is in this context, through innumerable species and over the expanse of time, that genuine altruistic emotions most likely emerged, giving divine Unlimited Love a place to begin work.

8

A CHRISTIAN ETHICAL PERSPECTIVE

ON LOVE'S EXTENSIVITY

ANALOGICAL THINKING frequently finds expression in metaphor, allegory, and parable. Such analogical reasoning is a powerful pedagogical tool, and is well suited for moral education. The familial analogy is like the air we breathe; it is so deeply embedded in the language and traditions of a community of interpretation that the user is not always aware of its centrality. We live most of our lives in various domestic and familial contexts, so drawing on the relevant set of analogies is as sensible for us as it was for Jesus of Nazareth.

For those who see nothing in human nature that in any way resonates with the *agape* or unlimited love of divine emotion, all analogies to human nature are distortions. While I have expressed doubts about analogies between human and divine nature, there is simply no other way to talk about divine love than by considering the highest forms of love that we know through nature. As I argued in the previous chapters, parental affection is the surest foundation of other-regarding love within nature and has a partial likeness to divine love. It is difficult not to detect some elements of the "self-emptying" love of God captured in the Greek word *kenosis* (Philippians 2:7) in the hearts of most parents. The Quaker theologian George F. R. Ellis defines this love in terms of a "deep morality," which is "a joyous, kind, and loving attitude that is willing to give up selfish desires and to make sacrifices on behalf of others."[1] This self-emptying love, argues Ellis, underlies and sustains the creation of an ordered universe. It also underlies and sustains new lives born into the world. Procreation is a participation in creation under similar circumstances of generativity.

One major objection to relying on parental love as a touch point for thinking about unlimited love is its admittedly insular focus, however much it is the evolutionary basis for more extensive altruistic love. One of the best examples of parental love unleashed from genealogical restraints is adoption, and this provides an opportunity to discuss the plasticity of this love as it moves outward to embrace the child who is unknown and genetically foreign. From this discussion we will move toward other aspects of the extension of parental love.

THE CREATIVITY OF ADOPTIVE LOVE

Many evolutionary biologists dismiss adoption because it undermines their theory that human altruism is determined by genetic interests. David P. Barash, with no significant evidence to support his proposition, treats adopting parents as psychologically confused by impulses left over from our hunter-gatherer past. "Each of us," he writes, "is a genetic slingshot, a catapult that shoots genes into the future." Barash acknowledges that we are unique in "being able to say no to various biological imperatives, breeding not the least." But the fact that we are capable of saying no does not mean that "we typically do so, or that it always feels good."[2] Barash does not think that, on a certain psychological level, adopting parents can possibly feel content: "After all, to adopt is to expend time and resources on behalf of someone unrelated to the adoptee." It may seem like adoption is "genuine altruism (that is, beneficence toward another without compensation of either kin selection or reciprocity)," but according to Barash this is not really true.[3] He offers as evidence the fact that adoption is not most people's first choice, in that they would rather have biological offspring. I must quickly interject, however, that while many parents adopt after they have tried unsuccessfully to procreate, there are also many parents with biological children who also adopt, and there are many who adopt children without making any prior effort to procreate. Such adopting parents may do so out of spiritual commitments to the ideal of *agape* or unlimited love.[4]

Yet Barash, who draws on none of the significant existing social scientific studies about the human experience of relinquishment and adoption, simply asserts that it can only be unsatisfactory because it flies in the face of his theoretical models. While adoption may satisfy the evolutionary desire to be a parent, he believes that it can never be ultimately fulfilling.

Moreover, in a remarkable exhibition of undocumented assertions, Barash argues that when adoption evolved in small hunter-gatherer groups, the adopted child was likely to have been a genetic relative, satisfying the demands of kin-altruism; in addition, it surely resulted in various forms of reciprocity from the group at large, satisfying the demands of "Tit for Tat." Adoption today, Barash concludes, is merely an evolutionary holdover. The problem with Barash's reasoning is that he knows nothing about the history and practice of adoption; if he did, he would realize that it does not fit his theoretical models.

Yet such reasoning contains a grain of truth. The genealogical family combines the bearing and rearing of the child in a manner that nature recommends. A married couple's wish to have their "own" child captures something fundamental.[5] There is no question that the sciences of evolutionary biology and evolutionary psychology, grounded in the biological investment of parents in continuing their genotype into future generations, support the fundamental nature of the drive to raise a child of one's own making. A mother and father see themselves in the child; correlatively, the child benefits from identifying his or her biological lineage, although this is hardly essential for a child's thriving. The successful practice of adoption, however, is proof that parents do transcend the "selfish gene" of the early-school evolutionary biologists, and that children do prosper without the narrative of a biological lineage.

The importance of connecting bearing and rearing is reflected in the fact that the adoptive parent-child relationship has been described in law with the terms "as-if-begotten" and "as-if-genealogical."[6] The adopted child is granted a new birth certificate. The adopting parents are even sometimes matched with the child's basic physiological features, enabling them to take on the characteristics of birth parents so the child will not "miss" a sense of biological ancestry. This mimicry, however, strikes me as biased against adoption, and is increasingly rare in adoption practice. Moreover, current practice is to inform the child of the adoption, which makes mimicry futile.[7] The majority of adoptive parents inform the child early, generally around age five, although some wait longer.[8] This mimicry also has the unfortunate result of implying that the genealogical family is the "real" one, thus encouraging the perception of the adoptive family as inferior to the biological one. Adoption does not need mimicry. Instead, it should be exalted for the salutary love that it manifests.

Adoption is a vivid counter-gene practice; it puts the almighty gene in its place. The gene has been described as "a cultural icon, a symbol, almost a magical force" and as the secular equivalent of the soul: "Fundamental to identity, DNA seems to explain individual differences, moral order, and human fate."[9] Sociologist Marque-Luisa Mirangoff has defined genetic welfare as a distinctive worldview that insists on degrees of genetic perfection, somewhat to the detriment of the social-welfare orientation that stresses environment and social intervention. People begin to see the world differently. "The emergence of Genetic Welfare," writes Mirangoff, "unlike the 'passionate movements' of the past, is a quiet revolution insinuating itself into everyday life in incremental fashion."[10]

Christian ethics is deeply appreciative of the birth ties between parent and child as a matter of natural law. However, it neither suggests pretending that the blood connection is there in cases of adoption, nor supposes that the adopted individual will necessarily need to search for his or her genealogical or supposedly "true" familial self-identity. Christianity challenges the assumption that the only real kinship is based on birth, biology, and blood. The ties of nature are important, but not absolute under the freedom of God. Families can be built as well as they can be begotten; every principle in action admits of some exception. Parents of adopted children lament the narrowness of a culture in which the adoptive parent is not considered the child's "real" parent.[11]

The Christian moral tradition has made a significant place for adoption that is a useful example of how parental love can be freed from biological constraints by spirituality and culture. The idea of adoption was contributed to Christian tradition by St. Paul. Five of Paul's texts mention adoption as a means of obtaining permanent enjoyment of an improved status as legal heir and having old debts canceled. The Pauline appropriation of a theology of adoption is exceedingly complex and beyond the scope of this discussion, except to note that it is clearly a part of his vision of Christian membership and salvation.[12] In essence, human beings are *not* the children of God, and cannot achieve this status no matter how hard they might strive to do good. Jesus, as the Christ, however, is the Son of God, and God is prepared to adopt as children those who accept Christ. Adoption is thus at the very core of the Christian narrative of salvation by faith: Christians view themselves as adopted sons and daughters of God through faith in Christ. Cultural beliefs and symbols do influence the direction of our loves. In essence, Christianity is a religion of adoption that creates a culture in which parental love is widened beyond the purview of genetic essentialism.

Christianity is completely unique among world religions in making adoption such a central motif.

Over the course of western history this theological endorsement of adoption as a blessing and a good has spilled over into actual social practice at the familial level, although this is not an entirely consistent development. Kristin Gager, for example, argues that Christian leaders in France during the early medieval period opposed the practice of adoption because it allowed the continuation of pagan family cults.[13] At that time, the church used a ban on adoption to restrict the power of clans and lineages by limiting the opportunity for nonbiological heirs. This meant that more land would pass to the church by default.[14] This ban, however, was difficult to implement and its impact was limited. Under different circumstances and in different periods the church has been strongly accepting of adoption.

There has been some historical resistance to adoption in England. In post-Reformation England, needy children were placed under Poor Law guardians and apprenticed; nineteenth-century English practice stressed institutions, emigration, and foster placement. The English were very reluctant to allow adoption because they so overemphasized the centrality of blood lineage. Adoption was only legally sanctioned in England in 1926.[15]

Thus, I do not wish to argue that the strength of a theology of adoption translated with perfect consistency into a social institution across the scope of Christian history. Indeed, adoption can be relied on in dubious ways. For example, the "orphan trains" of the mid-nineteenth century took children from eastern cities to western states like Ohio and Michigan, even against the protests of biological parents. This policy was supported by many Protestants.[16] The Roman Catholic Church, in contrast, was unwilling to so casually break up biological families; instead, it developed innumerable children's institutions to provide temporary church-sponsored relinquishment centers for Catholic parents.[17]

In general, however, the implementation of adoption under the cultural canopy of the Christianized western world has been highly salutary and is an excellent example of human liberation from genetic determinism. Christians believed that God had "given up a child to them," a child sacrificed by his "natal" father. Converts all believed that they were adopted into the faith, sometimes setting aside hostile biological families. They were "provided with a 'birth' through baptism—a kind of rescue of abandoned children."[18] The convert had "spiritual parents" who took special care in nurturing his or her faith; these parents "took up" a child, much like adoption.

Early Christianity endorsed adoption theologically as a metaphor for salvation; although more historical study is needed, it was also endorsed in practice as a necessary response to human contingencies. In ancient Rome, relinquishment occurred under a mythological canopy: a statue of a she-wolf suckling the foundlings Romulus and Remus stood over the forum in Rome from the third century B.C., "conveying to Romans who passed under it, for many subsequent centuries, the potentially happy prospects for abandoned children."[19] The benefits of relinquishment were thus creatively ensconced in cultural symbol and ethos. With the development from Roman antiquity to Christianized Europe in the fourth and fifth centuries, a new sacred canopy captured the happy prospects of the relinquished infant and the acceptability of the birth mother's gift, which at that time took on deeper religious meaning. The Christian self-perception was that "they had been 'substituted'—not unlike an abandoned child—for the posterity of Abraham." Given the centrality of this theology to the post-Constantinian formation of western culture, "Christian literature was filled with positive and idealized images of adoption and of transference from natal families to happier and more loving adopted kin groups."[20]

The Christian tradition provides an impressive cultural umbrella for the principle of rearing the children one brings into the world. The fact that Christianity creates a sacred canopy for rearing one's birth children is important, since the stability of cultural life and the well-being of the vast majority of children are both secured by this moral imperative. Christianity also provides a sacred canopy for the adoptive family, and this is equally important. By this means, the Christian community legitimizes families that are created purely by *agape,* rather than begotten biologically. It also constructs a meaning system for those human parents who cannot provide for the children born to them and must therefore relinquish their infants. Christianity had discounted the importance of lineage and descent that is prominent, but not absolute, in Jewish identity.[21] For the act of relinquishing a child to be meaningful, it must be performed freely.

The impact of a religious system and culture on the practice of adoptive parenting suggests that human beings are not narrowly determined by their genes, and that parental love can transcend the biological parent-child axis in the practice of adoption. It transcends this kin axis in other, even wider ways as well.

THE MORAL LOGIC OF UNIVERSAL LOVE

Unlimited love requires love for every person without exception. How does one realize such extensive love when the evolutionary context for the development of compassionate love seems to be kin-altruism and group altruism? I do not think that unlimited love requires the abrogation of natural loves for the near and dear; rather, it permits a love for family and friends within a wider, overarching context of universal beneficence that places these special relationships under a higher ordering principle. Historically, this balancing of love for the nearest with love for the neediest falls under the rubric of the *ordo amoris*, or "the order of love." Western moral theological thought on the *ordo amoris* has always required a leaning toward the neediest that counters familial insularity or overindulgence of the nearest.

In a useful analysis of the ethics of Christian priorities, Garth L. Hallett, S. J., focuses on a classic case of a father's choice between providing his son with a college education or directing these considerable resources toward saving many people perishing in a famine. Both of these needs are significant. Moreover, for a Christian, these needs should really be in tension or even in conflict. Hallett has drawn on specifically Christian ethics, as well as moral philosophy, to conclude that "to the extent that he can, the father should give preference to the starving."[22] He rightly points out that Christian ethics looks toward the neighbor, who is all humanity, and values human life as a gift of God. He extols figures such as Suzie Valadez, known in south Texas as "Queen of the Dump," for all her work feeding, clothing, and caring for thousands of impoverished Mexicans living in the Ciudad Juarez garbage dump. Her own children "were required to make real sacrifices in the interests of their mother's cause and grew up with little beyond the mere necessities."[23] This remains the Christian ideal, Hallett argues, even for those who cannot implement it. I take no issue with him.

Hallett's argument on behalf of the neediest human beings is in accord with scripture and Christian tradition. But on the routine level of day-to-day family living, I would be interested in more details about the lives of Suzie Valadez's children. Exactly what needs and wants are these children being told to sacrifice? How are they responding? Do they help their mother in her work? Do they feel that she gives them sufficient spiritual and emotional support?

This problem of ordering love is emotionally, psychologically, physically, financially, and spiritually complex. Hallett thinks that our children can learn much about ethics by seeing this sort of behavior and experiencing certain inconveniences.[24] Hallett reassures those of us with a complex family life that we are right to struggle, a belief that is consistent with the New Testament, the church fathers, Thomism, and good moral theory.[25] Evangelical writer Rodney Clapp is similarly reassuring when he contrasts the privatized and secularized middle class family with the not-so-private households of the truly Christian home—which he describes as a "mission base," consistent with both the New Testament, the early church, and the wider Christian tradition.[26] Every Christian family, even young children, must avoid the familial narcissism that focuses on constant emotional gratification and dyadic intimacy, consumerism, and material comfort; instead, the moral authority of the Christian tradition must be accepted in its command to "love thy neighbor as thyself"—the neighbor who is everyone, but especially the one most in need.

The distinctive and justifying elements in the Christian family are imperiled without active engagement in the community of faith and its service to the world. Perhaps a focus on service of the neediest would lower the levels of conflict leading to divorce in the Christian community; spouses would realize that their ideal of perfect emotional intimacy, consistent with the psychology of self-realization, should be replaced by an ethos of service to the world, which brings its own kind of more lasting union. Ultimately, Christian spouses and their children must be brought into greater intimacy through the spiritual harmony of purpose that emerges from the challenges of serving the world. Spouses and children then become servants to one another and to the world.[27]

The moral issue of ordering loyalties to biological family members, church community, and all humanity remains central to Christian ethics and to all moral thought. It has been restated by the contemporary Roman

Catholic ethicist Louis Janssens: "But at every moment our particular action can only benefit some, e.g., ourselves, a neighbor, a certain group. Why do we act for the well-being of this person or this group rather than for the advantage of others (the classical problem of the *ordo caritatis*)?"[28] Stephen J. Pope has even criticized the Catholic tradition for recent neglect of the *ordo caritatis*.[29] I suspect this neglect is related to the fact that contemporary Catholic ethics focuses on issues of social justice and liberation, making the earlier attention to the familial sphere appear less significant. As Hallett contends, however, this focus makes a great deal of sense in light of the emerging prosperity of many American families and the absolute poverty rampant throughout many parts of the world.[30] The lively debate over priorities will continue; complacency in the spheres of either near or distant need must be avoided.

There are people who do good for near and dear ones exclusively, giving insufficient attention to strangers. They make an idol of their closed social system and express no solicitude for those outside of it; they deny the universal loyalty that is grounded in the Christian assumption that no one is outside God's solicitude. This is why the family *needs* the community of faith that is open and dedicated to all people—as in Josiah Royce's Beloved Community that saves us from limited loyalties and enmity.[31] Western monotheism requires religious and moral suspicion of narrow loyalties; the metaphorical use of family concepts, such as mother, father, brother, and sister, creates an important new universal ethics. The greatness of Christianity lies in this extension of easily understandable images of the family to all humanity.

Suspicions of the narcissistic and self-indulgent family must not hide the equally significant reverse problem: loving the neediest stranger (who is also neighbor) while ignoring the truly legitimate emotional and physical needs of near and dear ones, especially children. I surmise that this problem is actually exceedingly rare; after all, people who fail to serve the essential needs of their own children are probably not much concerned with the neediest on the other side of the world, or even just down the street. Perhaps a married saintly figure with a daughter, such as Gandhi, could be criticized for being such a harsh taskmaster—although his daughter did ultimately learn from him and rise to become India's Prime Minister. Suzie Valadez could be criticized for having her children do without certain amenities, but she is probably teaching them a valuable moral lesson.

It is difficult to attend to strangers in any substantive sense until one learns to love those who are both genuinely needy and near and dear. This is why the crucible of the family is classically considered the seedbed of virtue for children and parents. Christianity points toward love of neighbor in a manner consistent with impartiality, for all persons are children of the same heavenly Father. True to Christianity, Augustine emphasized that all people are to be loved, "but since you cannot be good to all, you are to pay special attention to those who, by the accidents of time, or place, or circumstance, are brought into closer connection with you."[32] As though by "a sort of lot," Augustine wrote, some people happen to be nearer to the moral agent than others. Because they are embodied and temporal creatures, human beings are simply unable to love all humanity, except in intention. In this, Augustine followed the Stoics, as did Aquinas, who provides an account of the order of love in his *Summa Theologiae*. Aquinas argues that we should love those who are connected by "natural origin" most, except when they are "an obstacle between us and God." We might well differ from his conclusion that we should love our parents more than our children, and our fathers more than our mothers because men provide the "active principle" in conception.[33] Nevertheless, Aquinas seriously believes that, due to human finitude, an ordering of love is essential to the moral life.

Hallett emphasizes that while innumerable Christian theologians of antiquity and the medieval period recognized an order of love (with some interesting variations in detail), we cannot simply take their endorsement of special duties to the near and dear as an endorsement of familial insularity. Citing an array of sources, Hallett focuses on their appeals to our common humanity (as members of one family all sprung from Adam) in order to fulfill Christ's mandate (especially Matthew 25:31–46) to serve the poor, limit the rightful accumulation of property, value simplicity as an antidote to greed, and forego inordinate family affections. He concludes as follows: "The Fathers, who favored sharing even basic necessities, would more likely praise than blame a parent who sacrificed his son's higher education and chose to feed the starving." A careful analysis of the writings of Aquinas indicates that "it might be permissible, indeed preferable, for the father to assist the starving rather than his son."[34]

My point is that Augustine, Aquinas, and others realized that some rough ordering of love must allow for special considerations with respect to the family. As Henry Sidgwick noted, we should not resolve "all virtue into universal and impartial Benevolence," as though from the viewpoint of the moral agent the well-being of any one person is "equally important with the equal happiness of any other, as an element of the total."[35] Ethical theory should never swing too far from thinkers such as Thomas Aquinas, Bishop Butler, Adam Ferguson, Adam Smith, and Sidgwick, who believed in "the Order in which Individuals are recommended by Nature to our care and attention."[36] It is reasonable to first meet the genuine needs of those closest to us for whom we are particularly responsible, for example, as parents. Everything turns, of course, on definitions of genuine need in contrast to mere wants—and wants are too easily wrongly identified with needs. We sometimes diminish the significance of another's purported needs because we are simply too oblivious or arrogant to appreciate them as such.

Simple moral formulas will not suffice. No one can say exactly how much the physician should sacrifice the well-being of spouse and children in order to serve until midnight in an overcrowded clinic day after day. It would be overly rigid to lay out specific rules. There is inevitable variation in the particular loves and correlative commitments that individuals develop within the sphere of proximity. As a general point, however, this sphere holds some degree of acceptable priority in the moral life as long as the focus is need rather than indulgence.

To ignore proximity is to ignore the everyday moral dilemmas that weigh heavily on the conscience of most people. In his discussion of neighbor love, the eighteenth-century Anglican moralist Joseph Butler emphasized the importance of removing "prejudices against public spirit." Butler was as suspicious of "private self-interest" as any impartialist. Yet, following both Stoic and Christian precedent (*philanthropia* and *agapè*), he does not consider beneficence a "blind propension," but one "directed by reason." He states that "the care of some persons, suppose children and families, is particularly committed to our charge by nature and Providence; as also that there are other circumstances, such as friendship or former obligations, which require that we do good to some, preferably to others."[37]

Any worthwhile spiritual or ethical tradition rejects the provision of benefits for those near and dear that go beyond needs while strangers suffer from injustice, insofar as their minimal human needs are ignored. In other words, a problem of injustice arises when special relations sap our resources and energies in order to meet endless wants, while we care little for the most needy and for all humanity. This is why religious and philosophical thinkers have sometimes seen "special" relationships as subversive of wider commitments. The priority sometimes afforded family relations has been associated with the perpetuation of "unjust property arrangements, and deeply entrenched, self-perpetuating inequality of opportunity."[38] Plato, in his *Republic*, argued for the elimination of family relations altogether among the Guardians; in his later dialogue, the *Laws*, he still advocated severe regulation of the family. Even in these cases, Plato is talking about an ideal society. In the first two books of his *Politics*, Aristotle retrieved family life as a good, and the Romans followed suit. Cicero referred to the family as the "seedbed of the state."[39]

Following Aristotle, Cicero, and Thomas Aquinas,[40] unlimited love takes the social value of family relations and obligations seriously with respect to meeting needs. Its philosophy of the self takes into account biologically-based roles and relations, and views the powerful duties that correlate with these roles as socially beneficial. Often, families provide a uniquely loving environment for their members, and can provide economic support that would otherwise come from the state. This is a social and individual good.

Ethics is more than what goes on between strangers, people we see once and will likely never see again. It is recognized that to take away all distinctions of preference based on family relations or loyalties resulting from relationships over time is to strip the moral domain of those areas that are especially altruistic and loving. In 1915, Max Scheler warned the social philosophers of his time, "Looking away from oneself is here mistaken for love!" Finally, "All love for a part of mankind—nation, family, individual—now appears as an unjust *deprivation* of what we owe only to the totality." Scheler contrasts "love of mankind" with "love of one's neighbor," which is directed personally at near and visible others. Ethics is not "only interested in the *sum total* of human individuals." Scheler concludes that modern ethics makes concern with proximate persons appear "*a priori* as a *deprivation* of the rights due to the wider circle."[41] From the point of view of unlimited love, there is every reason to care for those who have been thrust into one's proximity by biological process and by the temporal and physical realities of existence. Unlimited love does not require us to turn away from those for whom we are particularly responsible. Yet there are many who have set aside all biological and social ties to work exclusively on behalf of the needy. The world needs such practitioners to model the sacrifices that such love can sometimes inspire, and to make us think more honestly and seriously about how our compassionate love might be too narrowly focused in ways that we do not easily wish to acknowledge and correct.

THE SPIRIT OF UNIVERSAL LOVE

Parental love is a complex phenomenon. It is not unlimited because it sometimes grows intolerant and impatient, overbearing and overcontrolling, and even violent and abusive. There are doubtless some elements in this love that might be described as proprietary and narcissistic—"I love this child because he or she looks like me and, with luck, will pursue more or less the same success in life that I have pursued." Parental love may not appreciate the importance of differentiation in the life of the child, who must establish his or her own identity.

But there is no serious question that this imperfect human love, which may be more or less limited, is the basis for whatever capacities we have for deep universal compassion. It seems, judging from the dominance of the parental metaphor in both the Hebrew Bible and the New Testament, to represent the most orthodox and simultaneously the most suggestive heuristic key into what is the universe's greatest wonder and mystery. The metaphor in the monotheistic traditions has been almost entirely paternal. "God the Father" has at certain times in history made fathers godlike, and it has always diminished the significance of the maternal features of divine nature. The dominance of the "God the Father" metaphor is rightly scorned by feminists.

Love implies an interest in and a concern for another that sets aside recalcitrant and adversarial impulses. We humans are mammals who nestle and cuddle our young, and we feel affection and compassion for them. Like many other mammalian species, a law of self-sacrifice emerges whereby parents invest in their offspring and even lay down their lives for their children. This parental quality of sacrifice is clearly one of the reasons why mammals, and humans in particular, have succeeded. Spirituality is in large part the endeavor to extend such loving affections to all humanity, including the weak, and even one's enemies.

The term "spirituality" in today's world seems to mean anything one wants it to. But the truest meaning of spirituality, consistent with all great spiritual traditions, is the manifestation of ever deeper and wide-ranging love, excluding no one and reaching out to all people, whatever their conditions.

Within the monotheistic faiths, divine love is clearly parental, our love for God is filial, and our love for one another is brotherly and sisterly. Unlimited love is love for every person as a member of the one family of God.

Of all loves, the most faithful is parental love. Parents take special interest in and responsibility for even the most imperiled newborns and debilitated children. They keep an open door for their children even when adolescence makes this difficult and opportunities are squandered. They try to sustain and enhance life. Unlimited love resonates with our highest human knowledge of integrated fatherly and motherly love enlarged to universal dimensions. The Anglican thinker Dorothy L. Sayers notes that "in books and sermons we express the relation between God and mankind in terms of human parenthood." She continues: "When we use these expressions, we know perfectly well that they are metaphors and analogies; what is more, we know perfectly well where the metaphor begins and ends."[42] We know that God does not procreate in the same sense as humans do, and we are using the analogy with a kind and benevolent parent in mind rather than a careless, cruel, and injudicious one. The metaphor is limited to the best kinds of behavior on the part of parents, and nothing less. Even at this, we know that such a metaphor is somewhat anthropomorphic, yet we measure God by our own experience because we have no other experience from which to measure. All our thinking about God is analogical, as Thomas Aquinas underscored. Sayers wrote of this analogical theologizing: "We need not be surprised at this, still less suppose that because it is analogical it is therefore valueless or without any relation to truth. The fact is, that all language about everything is analogical; we think in a series of metaphors."[43]

What unlimited love we achieve in life is shaped by what we have learned from others through their generosity. The importance of modeling love and thereby teaching it to others cannot be overstated. Transposing this to a spiritual level, early Christians confessed, "We love because he loved us first" (1 John 4:19). It is natural for a child to respond in loving ways to parental love. So, too, spiritual persons, who perceive themselves as loved by God, respond in loving attitudes. As they gain insight into God's faithful love for every person without exception, they also wish to love and serve others in this same spirit.

It might be asked whether the literal experience of being a biological or adoptive parent is a necessary one in order to proceed onward to a deep love for all humanity. I would say, absolutely not. The evolutionary base of genuine love is primarily on the parent-child axis, and thus we can appreciate the selective pressures underlying the emergence of affective affirmation of a cherished other. But since the underlying groundwork for love has been established in human nature, it can be directly utilized in a variety of ways under the influence of moral idealism. Gandhi, for example, clearly set aside many of the delights of family life in order to accomplish his calling. Jesus of Nazareth seems to have overflowed with love, but had no children. Many people can be called to a life in which literal parenthood is sacrificed for the love of the neediest, as with Mother Teresa.

It might be asked at what point in history did the capacity for indiscriminate compassionate love manifest itself outside of the parent-child axis? This is unclear, although Sir John Eccles suggests that early in human prehistory (sixty thousand years ago) we see evidence of group altruism in the care of those incapacitated by severe injuries.[44] But tribal or group altruism does not approach the ideal of unlimited love for all humanity without exception. Here one must turn to the altruistic ideals embedded in the teachings of the great world religions.

The slow spiritual elevation of compassion lifted it above the narrow evolutionary confines of kin-altruism and from tribalism to love for humankind, and this is the end point of all worthwhile spirituality and of consequent moral consciousness.

CONCLUSIONS

This chapter is intended to be suggestive of the ways in which parental love can become widened in its scope. Under the right cultural influences, it is able to embrace the relinquished child in the practice of adoption, leaving aside all genetic concerns. It can be transposed into the higher key of love for all humanity through the teachings of worthwhile spiritual and religious traditions, which will often speak of love for humanity in parental metaphors and thereby appeal to parental affections because of their unique intensity. Yet even untransposed, parental love is part of a divine economy in which each of us can experience the unique attentiveness and undivided concern of a mother and father, from which we are able to make analogical leaps in imagining what divine love must be like.

9

FIVE DIMENSIONS OF UNLIMITED LOVE

IN PRACTICAL AND THEOLOGICAL CONTEXT

OTHER-REGARDING LOVE is usually considered the hallmark of both a moral and a spiritual way of life because it is perceived to be both a human virtue and an aspect of Ultimate Reality. As the distinguished scholar of world religions Huston Smith wrote, "religion begins . . . with renunciation of the ego's claims to finality."[1] So too does the moral life.

A person who lives in love has a lightheartedness and hopefulness that demonstrate internal freedom from self. Such freedom includes being lifted up above the weight of one's own cares; even care of self exists now primarily for the sake of others. A new self is released from an old self. The freedom of love is a kind of levity in which one no longer trudges through the world imprisoned by worldly cares or external expectations. Indeed, considerable numbers of people who determine that only a life of love makes any sense will gradually or sometimes abruptly break away from their past lives in order to pursue new callings with deeper meanings.

The inner freedom of love is linked with an interest in and a powerful sense of concern and responsibility for others. Such freedom includes inward levity—a certain whimsical quality that allows for celebration, playfulness, and joy; simultaneously, it is the most demanding form of freedom because it is so active in love. Thus, in many of the great servants of all humanity we see paradoxical combinations of gaiety and severity, spontaneity and hard discipline, levity and intensity. Love is manifest in "the wisdom of fools."

People with the levity of love have a strange attractive power, and are able to gather around them those who find in what they hear and see the love they long for. They draw people like magnets by the pure positive power of self-forgetfulness. It is, then, possible to speak of the power of love, but it is an attractive, rather than a coercive, power. Ultimately, what such leaders achieve for humanity is due less to teaching than to the manifestation of love in service.

Love is characterized by an inward freedom from contempt and hatred, which are deemed unacceptable inner states and condemned immediately as incipient violence. The thoughts, emotions, words, looks, and deeds of love are all essentially one. Internal states of contemplation, prayer, and meditation are important to a life of love insofar as they allow the self to gain mastery over negative emotions. Unkind thoughts, bitter words, and scornful looks are all contrary to the tender reverence of another person.

The freedom of those who live in love can shake our sense of normalcy. They can be quite heedless of self and unpredictable in their generosity toward everyone without exception. Because their vision of universal community is all-inclusive, intense tribal altruists detest them and, threatened by their vision of inclusivity and peace, will usually resist them. Many great leaders with an extensive love for humanity have been killed by those with extreme in-group loyalties. The mathematical language of game theory— "maximum rewards," "optimality," "utility," and "payoff"—just does not bind them. They are freely alive to the gift of life, and have a nonpossessive delight in and celebration of everyone.

In spiritual terms, such persons pursue a grammar of recomposition— that is, they seek a new spiritual intimacy with all humanity. They sense that Unlimited Love is the ultimate reality in the universe, freeing them from existential anxiety. Many of the cares and preoccupations of daily life are set aside, and they seem to live and breathe more for others than for self. Love guides and governs them in all circumstances. Perhaps this freedom makes them less prone to stress-induced illnesses, and perhaps they even live relatively long lives if they are not killed; these side-effects of love are not intended goals, however, and would give way in circumstances requiring the high risks that sometimes come with the territory of love.

At some level, people of love have the quality of the young child whose mother's expression of love conveys a sense that the universe can be fully trusted. As 1 John 4:18 reads, "Perfect love casts out fear." The Greek for

"perfect" also translates as "complete" or "fully developed." Love elimi-
nates fear because it includes a trust in God's goodness. God loves us perfectly. We are so perfectly safe in that love that we can be loving ourselves.

Freedom is linked, in the broadest sense, with spirituality. "Spirituality"
is a somewhat ambiguous term, but most writers mean by it *a transcendence
over self and a discovery of the meaning and worth of others for their own sake.* Thus,
spirituality is not necessarily associated with any set of religious beliefs or
experiences. In the simplest terms, spirituality simply refers to the direction
and ordering of one's loves, which define a person's essence. It is, then,
possible to speak of secular spirituality. But in its more typical and perennial expression, people speak of spirituality in the context of the idea that
our lives are neither full nor complete until they correspond with a presence in the universe that is higher than our own, and that presence is recognized as including, or even being entirely defined by, Unlimited Love.
Let us be perfectly clear: while spirituality has come to means all things to
all people, profound spirituality is centered on love. People who claim to be
spiritual sometimes fly ruthlessly into violence, are institutionally absorbed
in the struggle for power within rigid status hierarchies, or meditate for
stress reduction without any development of other-regarding virtues or
behaviors. True spirituality is characterized by only one thing: the manifestation of abiding love for others, which fully determines the meaning
and the goodness of our lives. Secular people who are deeply compassionate and other-regarding have achieved a high spirituality. Anyone who
so abides in love is *de facto* spiritual in the broadest sense of the term. The
extent to which we abide in love determines the freedom of our true being.

For those who associate unlimited love with divinity, the notion that
God is love implies we are not alone in living lives of love. A constant and
unchanging source of love in the universe can be drawn upon even in the
most troubled and stressful of times, making love truly of the *cosmos* and of
the *polis*, or cosmopolitan. Our lives are meaningful exactly to the extent
that they correspond with unlimited love. For most people across the ages,
a spiritual life is linked to perennial metaphysical assumptions about
Unlimited Love as the single ultimate reality, both in terms of an underlying
creative reality and a continuous self-emptying presence.[2] In communion
with this ultimate reality, widely referred to as God, a life in which darker
emotions of hate, vengeance, resentment, possessiveness, uncontrolled
anger, and impatience reign will give way to the more noble emotions of

generosity, forgiveness, compassion, and kindness. Since we are ultimately created by Unlimited Love, the purpose of life is to cooperate with that love, however imperfectly. This cooperation is the full measure of internal freedom.

Metaphysicians contend that love is an underlying reality that pervades the universe in various manifestations. A century ago, the philosopher Charles S. Pierce proclaimed "the great evolutionary agency of the universe to be Love." He distinguished this agency from *eros* or passion, and instead described it as a "cherishing love." Referring to the Gospel of John, Pierce wrote, "Nevertheless, the ontological gospeller, in whose days those views were familiar topics, made the One Supreme Being, by whom all things were made out of nothing, to be cherishing love." He also pointed out that "love is directed not to abstractions but to persons."[3]

In this chapter, I will discuss from theological and practical perspectives the nature of unlimited love with respect to the five dimensions suggested by Sorokin that I identified scientifically in chapter two: *intensity, extensivity, duration, purity, and adequacy.* No love can ever become high in any of these five dimensions if it is not free, and no love relies on external force, except in the most regrettable contexts such as the self-destroying psychotic, the violent fanatic, or the malicious. I will locate my discussion of these five dimensions of love within a set of unrelated practical contexts selected because they lend themselves to the clarification of those dimensions.

Intensity

Altruistic or other-regarding behavior can be routine and casual. Giving up one's seat on the bus in favor of an elderly person or giving a poor man a dollar does not require great effort. Low intensity altruism is also the foundation of respect for persons, which is the basis of common morality. An affirming smile and a kind greeting are small but highly significant expressions of respectful other-regarding behavior—the recovery of which is welcome as an antidote to the downward spiral of a culture that has witnessed the rise of rudeness.[4] Etiquette, however culturally relative, is an important form of other-regarding behavior without which the quality of societal life slips away.[5] Every word and behavior is a mode of symbolic expression through which we can convey respect for others. Physicians who are kind enough to offer a patient a glass of water, make the occasional home visit, or attend a memorial for the patient who has died express love in a manner that does not require intensity.

A life of love weaves together large and small acts constantly in all daily social contexts. The idea that the agent of love can behave bitterly at work but lovingly at home is generally false. Love is a virtue that inclines one to act consistently both around the clock and across all of life's domains; people of genuine character do not allow themselves to become bifurcated, acting callously in one context and generously in another.

There are circumstances, however, that call forth love in its most intensive expression. Martin Luther King, Jr., for example, conferred with his mentors before deciding to take on the cause of civil rights, knowing that he would be endangering his wife and children.[6] Dietrich Bonhoeffer, born in Breslau in 1906 and the son of a famous German psychiatrist, was studying in New York City at Union Theological Seminary when he left the safety of America to return to Germany and continue his public opposition to the Nazis. He was arrested in 1943, having been linked with a group of conspirators whose plot to assassinate Hitler failed. He was hanged in prison in 1945.[7] King prayed to God for "the strength to love," and Bonhoeffer struggled to accept "the cost of discipleship." Both are widely heralded for their courageous decisions. Although they were not seeking death, they both knew it was a possibility, and death eventually found them. Before they made their momentous and inspiring decisions, King and Bonhoeffer were men of considerable talent but otherwise ordinary. Every well-known exemplar of intense love in the form of compassion and service is a brave, ordinary person inspired to respond to difficult circumstances, whether racism, fanatic self-righteousness, or a thousand other incursions of evil.

The thousands of people who risked health and well-being to clean up after the World Trade Center attacks also rose to the occasion. Soon after the attacks on September 11, 2001, the Rev. Dr. Frederic B. Burnham of Trinity Church, located next to Ground Zero, received a request for research proposals from the Institute for Research on Unlimited Love. He sent me this email describing the intensity of love in the wake of 9/11:

> I was delighted to receive your email on proposals on unlimited love. I had no thought of seeking support for a project I have been working on, but by God's grace your email just dropped out of the sky.

As you may recall I am the Director of Trinity Institute, a program of continuing theological education sponsored by Trinity Church/Wall Street. I am also a historian of science particularly interested in chaos and complexity. Since September 11th I have been deeply involved in St. Paul's, a chapel of Trinity Church that has been turned into a 24/7 refuge for the relief workers at Ground Zero. It is an astounding ministry. It is a place which perfectly models unlimited love: "Unselfish delight in the well-being of others." Since day one I have also realized that the principles of chaos and complexity wonderfully illuminate the emergence of and continuous twists and turns in that spirit of unlimited love that fills St. Paul's. . . .

I am inviting you to come see this living laboratory for unlimited love . . . I am not only inviting you to take a field trip, but I believe I am offering you the opportunity to see a remarkable example of the human spirit at its best.

On February 12, 2002, I spent an afternoon at St. Paul's Chapel of Trinity Church, Wall Street. The chapel, built in 1766 and full of history, is perched precariously on the cliff-like border of the cavernous site of Ground Zero. Immediately after the events of September 11, ministers from Trinity Church, General Theological Seminary, and the Seaman's Church Institute joined other volunteers to launch a ministry at St. Paul's for the many rescue workers. Upon entering, one saw walls covered with artwork and messages from well-wishers around the world. There were cots here and there, although the rescue workers mostly slept on the pews. Soup, sandwiches, and snacks were available. Helpers came from all over the United States, although they only stayed for a few days to make room for others who wished to volunteer.

It would be impossible to exaggerate the amount of love energy I sensed that day at St. Paul's, even after nearly half a year of round-the-clock ministry. The Rev. Lyndon F. Harris, a doctoral student from General Theological Seminary and associate minister at St. Paul's, had been serving long hours at St. Paul's every day since September 12. The Rev. Harris is an ordinary man from South Carolina who responded selflessly to an unprecedented set of circumstances. I noticed that he knew all the rescue workers personally and they were constantly wanting to talk with him. He gave a lot of attention to the details of their care, including all the practicalities involved in making St. Paul's into their virtual home. Now that it was five months after 9/11, the Rev. Harris was finally about to take a couple of days off. He told me that while the work had been challenging, he saw this opportunity for service as a once-in-a-lifetime calling from God, and that he and his staff were privileged to see and hear and serve. All felt that serving in this ministry was an honor, and all imperiled their health to some degree in the process, especially early on when the air was thick with black dust and chemicals. Perhaps they had overcome the fear of death and were thus completely able to love. I felt that the events at St. Paul's, including the very survival of the building itself, were in the hands of divine love.

The palpable intensity of the love at St. Paul's Chapel stayed with me for days. It might be related to the radically disruptive influence of catastrophe as well as to the stark reality of profound human need. As existentialist theologians point out, human beings wish to protect themselves through the security of daily routines that provide order and control over existence. Routines and ordinary life goals, both of which create security and order, are lost when serious illnesses break into human experience. Fragile human beings are subject to critical and life-threatening contingencies over which they hold no ultimate control; thus, our routines are not as real as we imagine—and may even be illusory, while life goals are often superficial. It is at these critical life junctures that the spirituality of love becomes even more important in the lives of remarkably large numbers of people.

In such situations, human beings can discover their true selves through love, and they can reach higher toward unlimited love. They are freed of the egoistic illusion of self-centrality and self-control, and all the protective routines of everyday life have completely evaporated. The time comes to cease hiding behind routines and step out over the usual lines and boundaries to discover the significance of others. We become ourselves in these limit-breaking situations by a change in our consciousness of the significance of others.

The human creature seems to have a perennial sense of a high source of unlimited love, but never achieves it fully because of frailty and innumerable imperfections of the heart. We recognize our imperfections by humbly acknowledging that "God is love." In communion with unlimited love as a spiritual energy, love often seems to intensify, and we become more consistently the instruments of this love. Innumerable stories of people's lives being changed by a spiritual experience and turned toward other-regarding love are common knowledge. The question is whether this intensification has objective metaphysical groundings.

Is the universe more likely the result of pure chance or creative love in which we can participate? When a person loses the sense that the universe and life have a purpose in love, it is a serious matter. Religions have traditionally taught that such cosmic purpose does exist. Teilhard de Chardin saw the radical improbability of a line of development from the "stuff of the universe" to "the birth of thought" and the awareness of a "Great Presence." He conceived of this Presence as not being "up there," but as "up ahead," drawing the universe into the future toward unity. "A universal love," he argues, "is not only psychologically possible; it is the only complete and final way in which we are able to love."[8]

Some people do seem to manifest degrees of intense other-regarding love through their experiences. By participating to various degrees in God as Unlimited Love, we can affirm the value and even sacredness of the existence of all others with high intensity, leading to remarkable expressions of compassion, care, communion, and forgiveness. There is an emotional migration at the center of our being toward enhanced and more powerful love. For the vast majority of us who only participate in Unlimited Love to some very small extent, the fact that such love is a perceived reality in the universe provides immense solace and comfort. This spiritual intensification of love is widely documented. In Psalm 119, Judaism speaks of the *hesed* or "steadfast love" of God; Buddhism speaks of *karuna*, meaning compassionate love. One finds rough equivalents of the ideal of divine Unlimited Love across the major spiritual and religious traditions. Moreover, the consensus among religions about the various ways human beings can connect with this source of love needs to be better understood. How to best harness the power of Unlimited Love deserves further research and study.

Authentic religious experience, argued William James, is anything but self-preoccupying. The practical fruit of such experience must be heightened altruism. The "ripe fruits of religion" are universally understood in terms of saintliness, which includes these features: a sense of the existence of an Ideal Power, a self-surrender to its control, a sense of elation and freedom, and "a shifting of the emotional centre toward loving and harmonious affections, towards 'yes, yes,' and away from 'no,' where the claims of the non-ego are concerned." Religious experience, in relation to charity, is defined thus: "The shifting of the emotional centre brings, secondly, increased charity, tenderness, for fellow-creatures. The ordinary motives to antipathy, which usually set such close bounds to tenderness among human beings, are inhibited. The saint loves his enemies, and treats loathsome beggars as his brothers." Further, "Brotherly love would follow logically from the assurance of God's friendly presence, the notion of our brotherhood as men being an immediate inference from that of God's fatherhood of us all." In this experience of the religious propensity, James concludes, the "altruistic impulses" become more marked.[9] Yet so many of the everyday actions of unlimited love do not demand high intensity. The simple actions of common courtesy and the tones of kindness in speech or other expressions contribute immensely to the love energy of the social world. A minor act of random kindness can be a manifestation of underlying love. Some years ago, I was giving a speech for the Atlanta Chapter of the Alzheimer's Association at the Carter Center in Atlanta, Georgia. The director of the chapter was in the parking lot attempting to carry a stack of heavy papers from the trunk of her car. A kind voice asked: "Could I help you, ma'am?" There stood former President Jimmy Carter, who humbly proceeded to be of service. Unlimited love takes very ordinary forms and yet is always prepared should the extraordinary be needed. We should never lose confidence in the potential of love as a solution to even the most difficult of human circumstances, and as providing the opportunity to manifest the human capacity for resilience.

EXTENSIVITY

While I will treat extensivity here with regard to humanity only, it also pertains to the welfare of sentient nonhuman species. St. Jerome, for example, one of the four fathers of the Latin (western) church, is the subject of a legend in which he acted with kindness toward a lion by removing a thorn from its paw. Francis of Assisi exemplified joyful and respectful proximity to animals. One does not have to be an interspecies egalitarian or an animal rights advocate to acknowledge that other species, insofar as they are sentient or capable of experiencing pain, should be treated humanely. Indeed, if there are species in the universe with capacities higher than our own, one hopes that they will treat us lowly humans respectfully.

The ideal of "love of humanity" was first expressed in antiquity with the Greek word *philanthropia*. In the writings of the Greek physician Hippocrates, for example, we read, "Where there is love of humanity (*philanthropia*) there is also love of the art (*philotechnia*)." It is debated as to whether such love for humanity was intended to go beyond one's community. Scholars generally translate Greek *philanthropia* in terms of compassion and kindness, rather than wide extensivity, although some caution against trivializing this Greek medical principle for its purported narrowness. One Hippocratic precept is, "Where there is love of humanity there is love of medical science." The physician is to assist "even aliens who lack resources."[10] It is likely that the ideal of love for humanity was available in a very rudimentary form among the Greeks and Romans, if only for the reason that the word *philanthropia* was often used interchangeably with *agape*, the New Testament Greek word for love of humanity; most scholars concede, however, that Judeo-Christian theologians took the ideal to new levels.[11]

Within the Christian tradition, philanthropy was supplanted by the religious notion of *agape* love. In the words of historian Darrel W. Amundsen, "Christian philanthropy was the expression of *agape*, an unlimited, freely given, sacrificial love that was not dependent on the worthiness of its object, since it was the manifestation of the very nature of God, who is himself *agape* (1 John 4:8)." Amundsen continues, "It was incumbent upon all Christians to extend care to the needy, especially the sick. By late antiquity the care of the sick had become a highly organized activity under the supervision of the local bishop."[12] Hospitals were a direct outgrowth of Christian *agape*, which included a sensitivity to the sanctity of all life that constituted a moral revolution in western consciousness. After three centuries of persecution, Christians opened hospitals and made care of the sick an expression of love. They were unusually heroic in the leprosaria. It is argued that Christianity gave rise to a decisive change in attitude toward the sick, who now assumed even a preferential position.[13] In modern times, we see this attitude in Mother Teresa, who was awarded the Nobel Prize for Peace in 1979 for her exemplary love for the homeless and sick strangers on the streets of Calcutta. It has been suggested that contemporary moral philosophy could gain much through reflection on such examples of "compassion, generosity, and self-sacrifice."[14]

Extensivity pertains not only to overcoming barriers with race, class, ethnicity and belief, but also to overcoming stigmas against people with mental illness, dementia, retardation, and all illnesses that involve social biases.

Agape superceded *philanthropia* and *humanitas*. By associating love for humanity with the very nature of God, *agape* love added urgency, spiritual depth, and cosmic significance to what was only a half-born ideal in pagan antiquity. Greek *philanthropia* clearly lacked this sense of urgency and depth. Perhaps it was closer, in practice, to what we would nowadays refer to as *noblesse oblige*, in which the very privileged provide an occasional moment of casual generosity.

There are two key New Testament passages that have inspired innumerable people to widen the extensivity of love: the Good Samaritan parable (Luke 10:30–37) and the requirement that all Christians assist the sick disinterestedly (Matthew 25:31–46). It would be difficult to imagine any passages more influential than these on western cultural history. Here, extensive love is presented as the apex of the virtues. The emotions and actions of love are not merely "supererogatory," or optional idealistic additions to the morally required principle of "do no harm" that one need not take too seriously. The elevation of the spiritual and, consequently, moral life includes an enlivening of beneficent love that is essential rather than optional.

With regard to the Jewish and Christian ethical traditions, Christine D. Pohl describes how a hospitality that is open to the "alien" or stranger rests at the center of the moral life. In Judaism and Christianity, there is a clear connection between marginality and hospitality: the more an individual or group is marginalized and neglected, the more an ethics of hospitality requires that they be tended to. The "strange" are the poor, the weak, the different, the tribal or national outsider, and all others who are not the beneficiaries of in-group preferences. Contemporary communities of hospitality within faith-based contexts include Vanier's l'Arche, Millard Fuller's Habitat for Humanity, myriad Good Samaritan hospitals, and thousands of other associations and institutions that carry forward ancient prescriptions of love. Pohl writes, "God's guest list includes a disconcerting number of poor and broken people, those who appear to bring little to any gathering except their need."[15] It is not clear whether the commandment in Leviticus to "love your neighbor as yourself" (19:18) originally applied to Jews only or to non-Jews as well. It is clear in the writings of the rabbi-physician Moses Maimonides, however, that over time the commandment certainly came to be applied to non-Jews.

Under Unlimited Love all strangers are transposed into neighbors. Loving a stranger is a moral challenge since it lacks the encouraging expectations of reciprocity that reinforce self-giving in ongoing social relations, and because we may have nothing in common with the stranger—or even find ourselves deeply repulsed. It is a love that aims toward the good, whether psycho-physical or religious, of an unknown person simply because that person exists. The Good Samaritan reaches out to help the neighbor who is a stranger; this act demonstrates a moral idealism that those unable to transcend ethnic, religious, or familial circles cannot achieve. In the parable, there is no indication whatsoever that the Samaritan expected anything in return for his generosity. Indeed, the Samaritan story ends with his promise to return and repay the innkeeper for any expenses involved in care. As far as we know, the Samaritan had no past or future relationship with the innkeeper that would allow us to conclude that he was really only seeking reputational status in a "Tit for Tat" context.

The provision of food, shelter, medical care, and psychological support never depend on the recipient's holding the right beliefs or being in the correct state of soul. Agapists have served the needy in these various respects for centuries in unconditional ethical response to need, adding to the "spirit of beneficence" in western moral life.[16] Since all human beings are equally valued as children of God, Unlimited Love views them as worthy of protection from harm and of the provision of goods. There is no person who lacks this basic value, for it is bestowed by the Creator independent of any inherent capacities or talents. Human life is sacred because it is derived ultimately from the creativity of a loving God.[17] Indeed, theologians uniformly emphasize that *agape* love is grounded in part in the notion of absolute equality under God.[18] The friendship or *philia* of Greek Aristotelian ethics is set aside in the name of universal equal-regarding commitments. As the Anglican theologian Jeremy Taylor wrote four centuries ago, "When friendships were the noblest things in the world, charity was little."[19] Unlimited love is, above all, non-preferential.

This equality even extends to love for enemies: "Love your enemies and pray for those who persecute you" (Matthew 5:44); "Love your enemies, do good to those who hate you, bless those who curse you, pray for those who abuse you" (Luke 6:27–28). It is simply impermissible to close the door of love in the face of anyone. The enemy is purposefully not well defined in these passages. In every relationship of significance, there are times when one simply has to determine to accept periods of resentment and even bitterness, praying that the other might be changed by modeling love. There may be cases in which one intervenes to protect an innocent person, or even oneself—but such interventions are never to be implemented with an attitude of hatred or bellicosity. Theories of justifiable war may justify necessary defense of self and others, but they never justify the internal indignity of hatred, or the harming of noncombatants.[20] Love for enemies means that one does not permit oneself to fall into internal violence; for those who practice only nonviolent resistance, such as Martin Luther King or Gandhi, it means that one cannot inflict external violence either.[21]

Economic philanthropy in modern times was enlivened by and even created through the spirit of *agape*. Philanthropy, in its modern sense, means something more than it did in ancient times, for it has been quickened and deepened by the ideal of Unlimited Love. It combines the joy of other-regarding service with the moral alchemy of turning gold into something better. John D. Rockefeller made the connection between *agape* or unlimited love and philanthropy explicit. Mr. Rockefeller, a man from upstate New York of rural background and humble origins, created the Standard Oil Company based in Cleveland, Ohio, where he is buried. In 1906 he said, "I believe the power to make money is a gift from God—just as the instincts for art, music, literature, the doctor's talent, yours—to be developed and used to the best of our ability for the good of mankind."[22] Rockefeller emerged from a deeply spiritual Baptist background and understood his life, and all lives, as gifts from God. He wrote often on the centrality of love for one's neighbor, who he saw as anyone and everyone in need. He believed in the God-given essential worth of every human being and the capacity of each person to realize his or her own unique human potential. He had a folkloric sense of moral accountability for what was done with his wealth. For those who did not quite grasp his actions, which clearly defined a new level of philanthropic magnitude in the history of capitalism, he would answer with this justly famous phrase, "God gave

me my money." Of course, it is true that Rockefeller was an extremely
tough business competitor, and that he monopolized a whole industry for
years, but this does not mean that his actions throughout were not heavily
shaped by the ideal of God's love for all humanity.

Philanthropic luminaries typically draw on the ideas of stewardship and
human potential. "Stewardship" is a term that is based on Judaism and
Christianity. It refers to the viewpoint that we do not actually own or possess anything; our task is to be faithful stewards of what is ultimately God's
property, including our own bodies and lives. While wealth is not evil in
and of itself, it is evil for stewards to squander it. Possessions are illusory,
for anything we have is a gift from God to be used for higher purposes.
Thus, Christianity gave greater depth to the idea of *philanthropia* as it was
known in antiquity.

Philanthropy in its modern form is a creative example of the merging of
Greco-Roman and Judeo-Christian aspects of western tradition. The
legacy of Jewish philanthropy is as distinguished as Christian. Philan-
thropists who give away so much manifest extensive love, in the broadest
sense of service to humanity. Undoubtedly, some prominent philanthro-
pists are concerned with reputational gain and social status. But many also
display motivational purity, as exemplified by Millard Fuller, the successful
businessman who gave up all his wealth to found Habitat for Humanity in
his conviction that there shall be "no more shacks."[23] Fuller's vision of
equality and extensive love takes form in his belief that "all of God's people should at least have a simple, decent place to live."

Extensivity of love is the single most significant spiritual and moral
accomplishment of *Homo sapiens*. It underlies all the major Enlightenment
ethical theories from Kantianism to utilitarianism, however secularized
and distorted. It is the distinctive and unique contribution of spiritual
underpinnings to all that we rationally consider good and right. Extensive
love is the salt of the earth, and it has inspired remarkable achievements,
from the founding of the world's first hospitals or the modern-day Dame
Cicely Saunders's St. Christopher's Hospice, to the great philanthropies
and public charities of contemporary America. *Philanthropia* became extensively unlimited love under the sacred canopy of *agape*. As John D. Rockefeller, Jr., wrote as his tenth principle for the Rockefeller charities, "I
believe that love is the greatest thing in the world; that it alone can overcome hate; that right can and will triumph over might."

DURATION

Unlimited love endures and never ends. It is an underlying attunement of the emotional self that does not change in the way that fleeting passions or emotional "inflammations" do. It is such a fundamental affective attunement that it is unaffected by external circumstances and forms the basis of serenity. Thus such love is associated in various traditions with the influence of divine spirit. Søren Kierkegaard sharply distinguished *agape* or divine love from all human love because of its constancy.[24] The enduring quality of unlimited love either shapes and forms our human relationships, or we are left with pure chaos and brokenness. Duration is not a burden but a blessing, and one that generally confers a sense of well-being upon all.[25]

The enduring quality of unlimited love is related to its being a non-acquisitive inclination.[26] Acquisitive tendencies include desires for food, drink, and possessions, and lead to merely instrumental relations with others. A strictly acquisitive love is not really love at all; it implies indifference to the other's welfare except as a means to self-gratification—that is, as a means of acquiring a good for the self, with benevolence never a controlling motive.[27] Unlimited love is not based on some perceived positive attribute (property-based love) in the other, but on the value of being itself (existence-based love). Property-based love is reason dependent in that X can provide property-based reasons for loving Y. The attractive properties of the object account for the presence of love and determine its longevity.[28] Critics of a strictly property-based love believe that it leaves love too insecure, since the properties for which X loves Y may disappear, or X may no longer perceive them as attractive. They therefore introduce a love of bestowal, or a nonappraising love, except in the sense that all that exists is deemed good and worthy of grateful acceptance. Any enduring love goes beyond strictly appraisive categories.[29]

In general, love that does not endure is false. Our ideal of love is that it should last and be the foundation of permanence in human relations. Fleeting love does not create the sense of security that human attachment requires. There is no safe haven in such love. The travesty of fleeting love is as clear in its adverse impact on the development of a child as it is in the domain of conjugal intimacy.

Because clinical psychiatrists are close to the pulse of modern culture, their perspectives on the fleetingness and chaos of love in time are grounded in medical concern rather than abstraction. Psychiatrist Willard Gaylin describes the current American scenario of today bluntly:

> While the final score is not yet in, the results so far of this so-called "sexual revolution" are less than reassuring. The Freudian view of human behavior laid the positive groundwork for the liberation of the sexual aspirations of women from both an oppressive personal sense of guilt and the shame and humiliation of social stigmatiza-tion. But the only empirical results of that illegitimate offspring of Freudian philosophy, the sexual revolution, seem to be the spread of two sexually transmitted diseases, genital herpes and AIDS; an extraordinary rise in the incidence of cancer of the cervix; and a disastrous epidemic of teenage pregnancies.[30]

Gaylin, a romantic personalist with respect to philosophies of love, opposes "the trend towards highlighting the erotic aspects of sexual inti-macy and permitting the separation of sex from intimate social bonds."[31] Another psychiatrist, Paul R. Fleischman, writes that if sexual repression dominated the psychological landscape in Freud's Vienna, the current problem is quite the reverse. Fleischman's assessment of current clinical psychiatric practice is central to my own concerns about the perils of unstable love:

> Among the hurt and pained in need of help, who may suffer from broken marriages, fluctuating or fallen self-esteem, obsessive con-strictions, panicky attachments to parents, bewildering isolation, uncontrolled rages, and haunting depressions, the common denominator is an inability to transcend themselves with care and delight, to reach over and touch another heart.[32]

Fleischman's patients report that they suffer emotionally because they have assumed that well-being is associated with impermanent love. They then pursue such relationships, and suffer the consequences. Their experience may be summed up thus:

> The binding together, the touch of person to person, is sought concretely, rather than spiritually, and dyadically rather than com-munally. The substitution of sexuality for religious life constitutes one of the most prominent and pervasive elements of cultural pathology that a psychotherapist encounters.[33]

Many people seek to touch physically for the sake of sexual intimacy alone, failing to see physical touch as being expressive of a deeper spiritual meaning and enduring love. Enduring love is nowhere to be found.

Such concerns are prominent in current psychiatric circles. They go back at least to Rollo May's criticism of the depersonalizing tendencies of a culture obsessed with sex as a mechanical function and as the mandatory expression of any relationship worth mentioning. May considered the sexual libertarianism of the late 1960s to be a "new straight jacket." He argues that in so-called "free love" we do not learn to love, for such learning is work and requires long-term commitment so that relationships can serve as crucibles of moral development.[34] Feminists such as Catherine A. MacKinnon have played an important role in pointing out that the proclaimed freedoms in the sexual domain have masked wanton violence, date rapes, and other oppression of women.[35]

In the 1950s, C. S. Lewis rightly warned against the loss of lasting love and the ascendancy of mere sexual intimacy: "Poster after poster, film after film, novel after novel, associate the idea of sexual indulgence with the ideas of health, normality, youth, frankness, and good humour. Now this association is a lie." It is a lie, wrote Lewis, because sexual indulgence without commitment and steadfast love has always been associated with disease, deception, jealousies, and emotional pain. Lewis claimed that our society had lost sight of definitions of enduring love, and that the result was oppressive. He rejected the practice of sexual union when it is isolated "from all the other kinds of union which were intended to go along with it and make up the total union."[36]

Unlimited love saves romantic and sexual love from themselves. The honest romantics have always understood the fleeting quality of romantic love, which is clearly based on an assessment of certain desirable qualities in the beloved. It turns out, as Freud appreciated, that these assessments are often projected by the lover, and are not real in the beloved; they often result in immense melancholy that the more reasonable among us would rather avoid. One need not go so far as Plato, who deemed romantic love "a grave mental disease," to recognize that on the scale of endurance it is not terribly high. The *eros* of romantic love requires the emergence of *agape* if it is to last beyond the inevitable loss of degrees of infatuation and passion.

Romantic love, however, is not simply a cultural creation from the early Middle Ages, nor is it a peculiar obsession of modern times. Evolutionary psychologists are probably right: this capacity to project various perfections onto an otherwise quite normal object is a deeply engrained reproductive trick of the species that is unlikely to be swept away by reason. It is absolutely perennial, evolved as an aspect of pair bonding, and is surely genetically and hormonally hard-wired into our human nature. Sometimes romantic love does last a lifetime and remains vibrant for decades, although it is more often the case that after a few years it begins to fade and must be transposed into something higher. I have no quarrel with romantic love; however, unless it is shaped within a context of genuine underlying other-regarding love, it will not sustain relationships in the long term. Moreover, passion can regrettably come to be felt as "more real than the world," undermining the meaningful institution of marriage or of other equally important relationships of lasting significance.[37]

In spiritual traditions, the blending of genuine other-regarding love with romantic and sexual love occurs through the vows and rights of passage associated with marriage. It is the vow made in the presence of God and community that invokes a higher love than *eros* to save *eros* from itself, as I have discussed elsewhere.[38] Freedom is the introduction of Unlimited Love into the domain of these two natural loves, thus turning them into something more and something better.

PURITY

Purity pertains to psychological or motivational freedom from acquisitive and calculating tendencies. "Pure love" has a long history in religious thought. For example, the seventeenth-century French mystic Madame Guyon challenged the dominant Roman Catholic theology of the time by arguing that the highest expression of love counts no cost, seeks no return, desires nothing, and finds perfect joy in self-giving. There is, she argued, "never enough of disinterestedness."[39] In the 1690s, Guyon was denounced by Jacques Bossuet, Bishop of Meaux, in his reaffirmations of natural self-love. The French theologian François de Fénelon came to Guyon's defense with his publication of the classic work *Dissertation on Pure Love*. This work had a powerful influence on the Quaker spirituality of love in Pennsylvania, and on the Boston Transcendentalists of the 1800s. Indeed, Fénelon's name is incised with the names of William E. Channing and Theodore Parker on the wall of the Boston Public Library. This history underscores the influence of European "pure love" theology on American religious life and theories of love. Just as in France debates over the possibility of "pure love" erupted between Fénelon and Bossuet (the famous *Querelle du pur amour*), so also did they unfold in America.[40] The debate is a perennial one, and it continues on in modern theology in a variety of contexts.

The key question in these ongoing debates is the extent to which the agent of love can or should become self-forgetful or "disinterested" in the self. Ann Douglas, a feminist historian of American religion, has argued that the ideal of self-forgetful love was viewed by nineteenth-century Protestantism as fitting only for women, who existed "for the world's comfort."[41] Valerie Saiving, a feminist theologian, has been especially critical of the notion of *agape* love because it has imposed on women an ideal of "the negation of self," when in fact their most common form of "sin" has been this negation of self that has precluded a fitting affirmation of self within the community of being.[42] I have been generally affirming of these and related views, arguing for more communitarian or mutualist understandings of *agape*.[43] In particular, the ideal of unselfishness has been confused with selflessness, or the doctrine of "no-self," and this has had adverse consequences, especially for those who have in various ways been oppressed. It is no surprise that African-American ideals of love, as expressed by figures such as Clarence Jordan and Martin Luther King, Jr., have consistently emphasized the inclusive mutuality of *agape*.[44]

While mutuality is without question a fitting ideal in the domain of love, it is never the indispensable condition for unlimited love or any love of an elevated type. Mutuality must be left to take care of itself. If it occurs, splendid, but the willingness to be an agent of love should never be hampered by the setting of such conditions. Unlimited love is deeper than *philia* and mutuality, and this is why it endures. It is purified of self-interest, even as it allows for the care of the self. To confuse *agape* with *philia* or mutuality is a serious error. What do we think, for example, of the person who is entirely engrossed in friendships but who is completely uncharitable toward non-friends? Unlimited love welcomes communion and the joy of mutual love, but such transcendent love does not require or even directly intend these goods. Imagine how impoverished the world would be if Millard Fuller or Mother Teresa required something of the people for whom they labored? Love responds to need. No good doctor makes his or her responsiveness to needy patients contingent on some calculated reciprocal benefit to self.

By "self-sacrifice" I mean the voluntary forgoing of what one considers a great personal good for the sake of the good of another. When we care for the destitute or the broken of mind, there can never be a requirement of mutuality. While mutuality is a good and we should always be open to it, the reality is that in a broken world of frail and imperiled beings, other-regarding love cannot be sustained on the basis of "Tit for Tat" reciprocities. When a child lies ill in the intensive care unit, it is only the needs of the child that matter; to the good parent the idea of mutuality is strictly irrelevant. When a family or professional caregiver looks after a person with advanced progressive dementia who can recognize no one but who needs much attention, there is little realistic value in an ethos of mutuality. The demands of mutuality are simply too limiting for this form of freer love.

This matter of radical self-sacrifice to the point of accepting death is very real in the lives of many people who exemplify significant degrees of unlimited love. The Christian faith captures the theme of radical self-sacrifice, including sacrifice of one's life, under the shadow of the cross. Suffering or self-sacrifice for their own sake would be instances of masochism—a grave sin. Some spiritual writers do express themselves hyperbolically in order to pull their readers in a counter-cultural direction. This is especially important in a culture such as our own, which has made the avoidance of pain and suffering an ultimate end and extrication from self-sacrifice almost a categorical imperative.

Christian orthodoxy quickly established the centrality of the cross for theological and moral meaning: "And being found in human form, he humbled himself and became obedient to the point of death—even death on a cross" (Philippians 2:7–8). Pelikan writes that the followers of Jesus quickly concluded "that he lived in order to die, that his death was not the interruption of his life at all but its ultimate purpose." Pelikan adds that this view, which Paul articulated, pervades the New Testament and early Christian literature.[45] As Victor Paul Furnish concludes regarding Paul's theology, "First of all, Christ's death is viewed as an integral part of God's plan and purpose."[46] Jesus died "at the appropriate time" (Romans 5:6); his act of obedience is contrasted with Adam's disobedience (Romans 5:19).

The primacy of the cross is so central to most Christian theology that it cannot help but be afforded primacy in Christian ethics. The cross was essential and necessary in response to God's will and call; by extension, the cross is essential and necessary in the Christian moral life, even if vicarious suffering for humanity is only possible for Christ himself. In a world hostile to God's purposes, the cross is sometimes unavoidable.

I do not think that unlimited love requires the active pursuit of radical self-sacrifice. I do, however, believe that anyone who chooses to live a life of love for every person without exception should anticipate encountering at some point a significant degree of self-sacrifice, especially when risking confrontation with clear evil. When they do, they should understand that this comes with the territory of unlimited love and defines its purity from self-interested motives.

ADEQUACY

So many expressions of love in the world are well intentioned but ineffective and unwise. Love can be successful or it can fail terribly, and one important measure of this is the efficiency of love. Of love, it can be said that "the road to hell is paved with good intentions." Any person who wishes to live a life of love must become competent to achieve fitting goals. Many exemplars of unlimited love are also very studious and learned because they desire to learn not for themselves, but for others. Most of us have no great admiration for selfish learners, but we hold in high esteem those who gain intellectual and practical knowledge in order to make the world a better place. Many patients in a hospital want a physician who is both compassionate and competent; if a choice must be made, they would prefer the physician who is competent and somewhat uncaring to the one who is wonderfully kind but incompetent.

How can love be wise and efficacious, rather than unproductive and even harmful? Adequacy includes care of the self as a means to serve others; it requires an ability to distinguish between fitting self-sacrifice and manipulative behavior on the part of others; it demands attention to the balance between love for all humanity, love for friends, and love for family members.[47] Any person trying to fulfill the commandments to love God and to love one's neighbor must first struggle with the inevitable pressures of the order of love. Embodied and finite creatures have real limits. In the end, there are no formulaic solutions to the order of love. While love for those near and dear can instruct us in the love for all humanity, it can also blind us to the neediest and result in unacceptable indulgences. The emotional, psychological, and physical uniqueness of each individual seems to suggest that the problem of ordering love will be acceptably resolved with considerable heterogeneity. It must, however, be firmly noted that spiritual traditions overwhelmingly challenge us to lean toward all humanity and away from those relationships that would require inattentiveness to the human family. Inevitably, love for all humanity requires certain sacrifices in the inner circles of life in order to serve the outer.

People of love are usually remarkably creative in what they do for others. Whether the other is an infant with disabilities, an adolescent without hope, a young spouse on the verge of a painful divorce, or an older adult with fading powers, *love always tries to find a way to be successful*. Millard Fuller really understands how to build a house; Dame Cicely Saunders really understands the details of good hospice care; Jean Vanier really understands the experience of retardation. As one of the more balanced biographers of Mother Teresa writes, she was "adept at dealing with world leaders and financiers and had a canny instinct for publicity."[48]

Love wants to be adequate and successful, and the agent of love must carefully study how to implement love effectively in challenging circumstances. Inspired by the purposes of love, many have gone on to become doctors or nurses, social workers, and managers of charitable corporations and foundations. One cannot simply intend love and do whatever one wishes without a solid fund of knowledge. True lovers of humanity pursue learning objectives that are deemed necessary to serve others well. Many become experts in their various fields. It is possible that people of love will earn a great deal of money to support effective philanthropy.

Unlimited love, then, is not "soft" or ineffective or incompetent, but rather learns all things deemed necessary to love well. People of love are willing to take on unwieldy administrative tasks in order to build lasting institutions; they are ready to take on multiple tasks even when learning curves are steep. People of love are not caught up in some mystical isolationism, but rather implement love effectively in the world.

CONCLUSIONS

Very few of us achieve high ratings on all five dimensions of love. Perfect unlimited love would be highly intense, fully extensive, unconditionally enduring, pure of selfishness, and extremely adequate or wise. People who achieve reasonably high ratings in all or most of these dimensions are exemplary, and should be studied scientifically. The more we can understand about such individuals, the easier it will be to encourage future generations to live relatively loving and, therefore, meaningful lives. The scientific study of such persons cannot capture everything about them, but it can begin to point us in certain directions with regard to human development, psychology, and pedagogy. We understand so much from a scientific perspective about human deficits such as antisocial behaviors and mental illness, and we apply every empirical method from brain scans to psycho-histories to improve our knowledge. The more we learn, the more we can hope to correct such problems in the future.

But we can also use scientific methods to study the positive side of human behavior, and there is nothing deemed more positive than altruism, love, compassion, service, forgiveness, and the like. It is time to start a new interdisciplinary field for the study of love and Unlimited Love, engaging great minds and turning the human future away from endless acrimony, hatred, and violence. Indeed, we have no alternative. We must discover ways to tap the treasure within us, as all spiritual and religious traditions have taught, for "the kingdom of heaven is within you."

The call to discover the perennial power of love that lies within us and the universe is more important than all the biotechnical schemes to extend the human life span or enhance some facet of human nature. Without love, no human accomplishments can be meaningful or lasting. How can love become more extensive and enduring? How can we raise our children to be better human beings than we are? How can we create a future that moves forward beyond divisive and violent insular altruism to universal altruism, in which tribalistic barriers between "them" and "us" are no longer meaningful or deemed valid? These are the questions any study of unlimited love must address.

PART THREE
DEVELOPING A SCIENTIFIC FIELD

10

THE SCIENTIFIC FIELD

OF UNLIMITED LOVE

THIS CHAPTER attempts to formulate a scientific field, primarily by asking questions. The interface of science and religion needs rigorous attention so that our knowledge may progress and our capacities for unlimited love be better understood and manifested. The science of love and of unlimited love is a bit like the science of the mind—cognitive science. The human mind is indeed mysterious and we may never understand it fully. But we nevertheless bring to the field of cognitive science a great many disciplines, including sociology, neuroscience, psychology, linguistics, mathematical models, computer science, artificial intelligence, anthropology, philosophy, and religious thought. Someday we may solve the riddle of mind, just as someday we may solve the riddle of understanding what makes some people so unselfish and loving, or deepen our insight into the infinite or unlimited love of God.

Science has a great deal to offer. For example, positive psychologist Michael McCullough developed a questionnaire to measure gratefulness as a personality trait. He found that this trait was positively associated with the frequency of engaging in altruistic behaviors such as tutoring, donating money, and volunteering.[1] Gratitude is an experience of awe and delight in the universe and in life. As Robert A. Emmons and Joanna Hill write, "Gratitude is a sense of thankfulness and joy in response to receiving—whether the gift be deserved or not, whether it is a concrete object or an abstract gesture of kindness."[2] Gratitude is a key feature of spiritual traditions that speak of the gift of life and of creation, and it is an affective state that is deliberately practiced and encouraged through worship, prayer, and scripture. The Psalms are filled with exhortations to gratitude, as is the Qur'an. Many definitions of love include gratitude as a key element. Love perhaps begins with thankfulness for the existence of the other.

There is much to learn about unlimited love and its power to effect change. Many of the better existing studies are remarkable. For example, in this now quite famous study, research subjects watched a documentary film about the work of Mother Teresa. Salivary immunoglobulin A (S-IgA) concentration, a marker for immune function, rose significantly and remained high for an hour after the film ended in those subjects who took part in an exercise in which they "recalled times in their lives when they had loved or been loved." The study concludes that dwelling on love appears to be important for strengthening this aspect of the immune system.[3] While the epidemiology of love is not a well-established field of inquiry, there are a number of other studies suggesting that altruistic love and compassion are salutary with regard to well-being and health. Yet it is fair to conclude that this area of study, which looks at the effects of compassionate love on health and longevity in both the giver and the recipient of love, is only just getting started.[4] I would be willing to hypothesize that compassionate love on the part of a health professional or family caregiver would enhance therapeutic efficacy.

In a well-established set of studies, brief intervention contacts between problem drinkers and highly empathic counselors have been shown to be highly effective. The empathy of the counselor clearly enhanced the long-term outcome of interventions. Brief counseling, even a single session, is as effective as longer courses of therapy so long as the counselor manifests high empathy levels. W. R. Miller concludes, after an exhaustive presentation of his lifetime of investigation in this area, that absolutely accepting love is able to evoke change in addictive behaviors.[5]

These and other studies suggest that human beings seek loving-kindness and prosper when they receive it. This should come as no surprise. At birth, babies are looking for a kind face, and they can discriminate between emotions quite quickly. Facial expressions of basic emotions are universal across all cultures. It is usually a mother's face that is found. If that face is smiling, then the baby feels comfort and safety; if that face is apprehensive, the baby feels fear. The brain takes shape in the early years of infancy and childhood in an emotional matrix of attachment. Throughout our lives we are seeking warm, generous, affectively affirming love. This is why love has so much power to transform and heal us, and why we delight in its presence. Love creates a world of safety and joy in which new attachments are formed that contribute to well-being.[6] The human need to be loved in a fully accepting manner is a universal phenomenon, and the consequences of love deprivation are significant and even severe. Unlimited love soothes and comforts the very essence of our being with feelings of homecoming, healing, and cleansing of negative affect.

It appears that giving love is also necessary and salutary. For example, many older adults suffer loss of purpose in life, which is associated with diminished mental health and depression; purpose in life has been linked to optimism and the absence of depression.[7] Older adults who volunteer experience increased life satisfaction and better perceived health.[8] Volunteerism seems to promote physical health and mental health.[9] These findings are interesting, to be sure, but are only representative of a much wider literature that suggests to many that compassionate love, altruistic behavior, and service probably run more with the grain of human nature and fulfillment than against it. Yet most of the science lies in the future, and I believe it will continue to show the wisdom of perennial teachings about love for the neighbor, who is everyone.

We know far too little. Is other-regarding love in some sense "in our genes"? Is it in some potential form in our neural-emotional inventory? Can it be measured? Do different groups or individuals have differing love potentialities? Is there a way in which love can be reinforced and strengthened? Is there some age in the life cycle when this potentiality can or must be captured?

In late January 2002, the Institute for Research on Unlimited Love put out a Request for Proposals that reached more than one hundred thousand leading researchers. Our goal was to stimulate research into loving behavior across all the scientific disciplines by initiating a new field of study among a group of highly-regarded researchers carefully selected from a large pool of potential respondents. I wish to offer here an abbreviated version of the Request for Proposals, and then turn to broader questions that need answering in order for us to make real progress in our understanding of altruism and love.

INSTITUTE FOR RESEARCH ON UNLIMITED LOVE
REQUEST FOR PROPOSALS

Introduction and Background

The Institute for Research on Unlimited Love is devoted to the scientific understanding of other-regarding love in all its manifestations from compassion and kindness to altruism and volunteerism. It seeks to encourage, support, and disseminate high-level scientific research on unlimited love so that we might better understand our human capacities and potential. The Institute also fosters dialogue among researchers, educators, and scholars in the humanities, theologians, exemplary practitioners, and the professions in a way that enhances our common understanding based on significant scientific findings.

The following is a broad definition of Unlimited Love: *The essence of love is to affectively affirm as well as to unselfishly delight in the well-being of others, and to engage in acts of care and service on their behalf; unlimited love extends this love to all others without exception in a manner that is enduring and constant. Widely considered the highest form of virtue, unlimited love is often deemed a creative presence underlying all of reality. Such love acknowledges for all humanity the absolutely full significance that, because of egoism, hatred, greed, and group conflict we otherwise acknowledge only for ourselves or for those closest to us.*

Rigorous scientific methods in numerous disciplines ranging from neuroscience to psychology are typically applied to the analysis of negative behaviors and psychological conditions. These methods, however, can and should be equally applied to the creative capacities for love, compassion, and altruism. (See S. G. Post, L. G. Underwood, J. P. Schloss, W. B. Hurlbut, eds., *Altruism and Altruistic Love: Science, Philosophy, & Religion in Dialogue* [Oxford University Press, 2002]; see also Pitirim A. Sorokin, *The Ways and Power of Love: Types, Factors, and Techniques of Moral Transformation*, reprinted from the 1954 original with a foreword by Stephen G. Post [Templeton Foundation Press, 2002].)

Unlimited Love is a principle affirmed universally by great traditions and leaders: "We can do no great things—only small things with great love" (Mother Teresa); "Kind speech and forgiveness is better than alms followed by injury" (Qur'an); "You shall love the alien as yourself" (Leviticus); "Love cures people—both the ones who give it and the ones who receive it" (Dr. Karl Menninger); "God is love" (1 John); "Sooner or later, all the peoples of the world will have to discover a way to live together in peace, and thereby transform this pending cosmic elegy into a creative psalm of brotherhood. If this is to be achieved, man must evolve for all human conflict a method which rejects revenge, aggression, and retaliation. The foundation of such a method is love" (Rev. Dr. Martin Luther King, Jr.).

Goals

✦ To better understand the human potential for loving and other-regarding emotion and behavior through studies from all scientific disciplines, including human development, epidemiology and health care, neurobiology and neuroscience, positive psychology, sociology, and evolutionary biology, as well as anthropology, political science, economics, and education.

✦ To better understand how the expression of Unlimited Love in society might be fostered, including attention to the roles of education, media, and spiritual-religious beliefs and practices.

✦ To promote widespread dialogue on the empirical, theoretical, practical, and socially beneficial dimensions of exemplary lives of service and love.

Research Program Areas and Sample Questions

Research proposals are welcome from all scientific disciplines. The following six research program areas have been developed by a team of research consultants who have written extensive white papers (visit www.unlimited loveinstitute.org, click on *publications*, and then on *research white papers*). These papers include details about the significance of the research area, history of existing research, methodological considerations, annotated bibliographies, and further elaboration on the sample questions listed below. Interdisciplinary applications that pertain to several program areas are also welcome. *All applicants are strongly encouraged to read the white papers before applying.* The sample questions below are by no means exhaustive.

(1) Human Development

This research area will focus on analyzing and synthesizing biological, psychological, sociological, spiritual, and religious aspects of the human developmental trajectory that may serve to foster unlimited love and altruism.

What are the variables in parent (mother)-child attunement and attachment that produce a gradient toward healthy and even extraordinary other (object) love as opposed to a gradient toward narcissism and even sociopathy? Another way to put the question is: How does parental love received become transduced in the child to the capacity to give love?

Do those high on altruism or empathy scales do better at choosing "attachment solutions" during separation stress or challenge?

Is the generativity that develops in later adult life on a continuum of human development with other-regarding love and Unlimited Love?

Is there anything we can learn about the human development of altruistic or unlimited love capacity from a greater understanding of the development of cluster B personalities built on the core of entitlement?

What are the neuropsychiatric elements of the human development of love?

What can be learned about the human development of Unlimited Love from "religious altruism" in the medical context?

What do we know about how we can teach altruistic behavior and empathy?

Is there a genetics of love?

How powerful are the effects of religious experience, belief, and/or behavior on the capacity to love in an exemplary way?

Do conditions of "brokenness" free us from inauthentic or routinized existence and provoke a response of other-regarding love?

(2) Public Health and Medicine

Cross-fertilization of the fields of positive psychology and social epidemiology is valued. Among "positive" psychosocial constructs, giving and receiving love has received preliminary scholarly attention requisite to furthering epidemiologic research. Mid-range theories of love have been proposed by psychologists; scales and indices have been developed and validated. Most of this work, though, has focused on romantic attachments, in keeping with the emphasis among researchers on the psychology of love. Broader multidimensional theories, such as the sophisticated taxonomy of sociologist Pitirim Sorokin, have been few. Sorokin conceived of love as comprising seven "aspects" (religious, ethical, ontological, physical, biological, psychological, social). Epidemiologic, clinical, basic-science, psychophysiological, and psychometric research are all viable approaches.

Is love a protective factor against morbidity and mortality?

Does love promote health, psychological well-being, and high-level wellness?

Does love exhibit therapeutic efficacy?

Is love a salutary component of the patient-provider relationship?

What are the physiological mediators of the love-health and love-healing relationships?

What are the psychosocial mediators of the love-health and love-healing relationships?

Are there psychophysiological correlates of giving or receiving love?

What can non-Western and unconventional perspectives on mind-body connections tell us about love?

Can new assessment instruments for love be developed and validated?

What are the most promising theoretical and conceptual models of love?

(3) Approaches to Defining Mechanisms
by Which Altruistic Love Affects Health

The chain of events leading from negative emotional responses to external events to impact on disease has been well worked out. There has been less research into the application of the same model to studying the effects of positive emotional responses on health. The same approaches and standards that have been used in the stress literature can be applied to the study of the chain of events leading from altruistic love to beneficial effects on health. It is possible that the mechanism by which altruistic love affects health occurs through blocking or attenuating the stress response, or through activating positive neurotransmitter pathways in the brain.

What can we learn about the neurobiology of altruistic love?

What are the neurohormonal and neurotransmitter transduction mechanisms associated with altruistic love?

Does altruistic love positively affect specific aspects of immune cell function or other measurable elements of health or disease? If so, are these effects mediated through neurophysiological pathways, neurotransmitters, neuropeptides, and neurohormones activated by altruistic love?

How do learning, memory, and early maternal environmental factors impact the development of altruistic love behaviors?

(4) Other-Regarding Virtues

Recent advances in the social sciences point to a variety of character traits or dispositions that may predispose people to positive interpersonal relationships. Some philosophers have referred to these traits as "virtues"— i.e., individual dispositions that prepare people for success in the social realm. Traits such as trust, generosity, faith, empathy, kindness, gratitude, forgiveness, and honesty are traits that one would ascribe to individual persons to varying degrees. Society and its institutions socialize people toward the acquisition and expression of these virtues because they are presumed to enhance helping behavior. How are these *other-regarding virtues* connected to personality and behavior, and to mental, physical, and social well-being? Religions and spiritualities provide many people with social and psychological resources that encourage virtues such as love. Proposals for studies are welcomed that investigate how this takes place—e.g., the extent to which spiritual moments (e.g., mystical states, peak experiences, experiences of awe or reverence) produce loving motivations and behavior.

What are the relationships between religion/spirituality and the other-regarding virtues?

What are the sociological and social-psychological conditions that foster other-regarding virtues and behavior?

What are the motivations behind other-regarding virtues and their implications?

How can other-regarding virtues be measured beyond standard self-report measures?

What action tendencies accompany other-regarding emotional states?

To what extent can other-regarding virtues be viewed as components of personality?

Are the other-regarding virtues related to health and well-being?

By what physiological or psychological pathways do other-regarding virtues influence health?

How are the other-regarding virtues related to social and interpersonal outcomes?

How can other-regarding virtues be promoted and taught?

(5) Evolutionary Perspectives on Other-Regard

Evolutionary biology is an important discipline for the Institute, both because Darwinian theory has profoundly influenced attitudes toward sacrificial love, and also because altruism is a central theoretical issue in the evolutionary explanation of social behavior. In recent decades renewed attention has been given to studies of altruism through the development or extension of new approaches. These include theories of kin and sexual selection, direct and indirect reciprocity, costly signaling and self-deception, pleiotropy, game theory, multi-level selection, and co-evolution or memetics. The Institute welcomes proposals in any area of evolutionary theory, but particularly invites studies that (a) pursue integration or comparative evaluation of contrasting theoretical approaches, and (b) are interdisciplinary and/or strongly empirical where appropriate.

To what extent are differences in other-regarding attitudes and behaviors (a) heritable or (b) associated with fitness differentials?

To what extent are kin selection and reciprocal altruism—and the associated familial and social affections—necessary or sufficient substrates for the development of deeply caring, richly benevolent other-regard?

To what extent is altruism a sexually selected trait, and under what conditions does it function positively or negatively in mate recruitment?

Do patterns in moral and altruistic behavior conform to expectations of indirect reciprocity, costly signaling, or self-deception theory?

How extendable is the influence of group selection, and must genuinely sacrificial other-regard inevitably come at the cost of intensely exclusionary disregard or hostility?

What social conditions mediate the expression of altruism?

How can non-adaptationist approaches (pleiotropic emergence, memetic transmission) be tested empirically and how might they contribute to understanding the cultivation of altruism?

Can game theoretic models of cooperativity be extended to account for genuine sacrifice by reformulation in light of group-level or hierarchical selection?

How can animal models contribute to our understanding of human altruistic behavior or its underlying affective qualities?

What is the relationship between organismal well-being and altruistic behaviors and disposition, i.e., is sacrificial love an internalized adaptation to life in groups?

How can religious beliefs, experiences, and institutions be understood in terms of evolutionary (genetic selection) or co-evolutionary (memetic transmission) processes, and in what ways are they biologically adaptive and/or altruism promoting?

(6) The Sociological Study of Faith-based Communities
and Their Activities in Relation to the Spiritual Ideal of Unlimited Love

Some would argue that the ills of contemporary life are largely traceable to a deficiency of altruistic behavior. Greed and the self-seeking nature of a consumer culture are considered responsible for the deterioration of social bonds that once kept life more humane. Americans from across the ideological spectrum have argued for the significance of civil society as an overlooked, yet integral, part of a functioning, healthy republic. As concerns about the level of civility and social capital in this country have grown, scholars have become increasingly interested in unraveling the very ingredients that define and produce civility and social capital. Many with an interest in civil society have been particularly interested in the question of what role faith-based organizations may play in countering the effects of greed and narcissistic behavior, thereby contributing to a more civil society in which altruistic behavior and other-directed love are more commonplace. Since many methodological and theoretical cross-cutting issues are relevant to this understudied topic, multidisciplinary projects are strongly encouraged. Interested scholars from sociology, political science, public administration, economics, and other relevant social science disciplines should consider applying.

Are faith-based organizations more effective than their secular counterparts in addressing various social problems?

Preliminary research seems to indicate that faith-based organizations are more effective in providing social services than secular or governmental counterparts. What it is that makes these initiatives more effective?

What are the most important variables influencing a person's decision to become a volunteer—and why?

How have faith-based organizations been so successful in mobilizing and sustaining so many volunteers?

What are the political, administrative, and legal conditions under which organizational cultures of principled agents exist and persist?

What are the most important determinates of civic engagement and participation?

How can religion or religiously motivated workers and organizations combat antisocial and egotistical behavior?

How can religion or religiously motivated workers and organizations promote pro-social behavior?

What is the extent of other-directed love dispensed by faith-based organizations?

Can faith-based organizations and government work together to combat social ills?

Does the potential for altruism or unlimited love exist in all people?

FURTHER ELABORATION OF QUESTIONS

In the above Request for Proposals the reader can see the seriousness of the questions raised and the potential scientific approaches to answering them. But the questions listed above are by no means exhaustive. We now turn to other questions, articulated with the help of Lynn G. Underwood. How can we better understand human loving motivations and actions, with a focus on all that these involve evolutionarily, genetically, developmentally, neurologically, socially, emotionally, theologically, and conceptually? How might such motivations and actions be fostered?

The empirical study of unlimited love can be organized into seven major areas: *spirituality/religion; ethics/philosophy; biology; psychology/human development; education; anthropology/sociology/political science/economics; medical.* There may well be better ways to group these disciplines, and there may be additional disciplines worth listing, but these are a good place to begin.

Spirituality/Religion
On this axis, Unlimited Love is identified with divinity: "God is love, and those who abide in love abide in God, and God abides in them," asserts the New Testament (1 John 4:16). The *Bhagavad Gita*, the *Dhammapada*, and the scriptures of virtually all the great world religions assert this identification with love as well, in their different terms. God is believed to be absolute Unlimited Love, although this need not exclude anger and judgment. This love is manifested in a way that does not ultimately discriminate against sinners. In many historical accounts, the inspired apostles of love—many great moral lights, founders of all genuine religions, and true sages, seers, and prophets—are also remembered as joyful. In spiritual and religious traditions, the life of altruistic and *agape* love has been understood to be a participation in divine love, replete with a sense of joy and universal extensiveness that goes beyond all intergroup conflict. This participation is shaped by the experience of prayer and meditation.

It would be good to encourage rigorous empirical studies into this phenomenon, as well as studies of the ways in which religious symbols, beliefs, stories, and rituals encourage agape love.

Some specific sample questions are:

> Do spiritual and religious experiences, beliefs, and practices influence behavior in the direction of altruistic, compassionate, and Unlimited Love? If so, when and how?

What specific spiritual practices (for example, types of prayer, meditation, silence, worship) might help to encourage altruistic love? How do these practices interact with the biological, social, and cultural substrate of the person?

Ethics/Philosophy

On the ethical plane, altruistic and unlimited love are associated with the unselfish affirmation, acceptance, and care of others for their own sake. Study of this association would include foci on the emergence and ascent of altruistic and unlimited love to constitute goodness itself in many significant moral traditions, and on how altruism is analyzed in contemporary philosophy with regard to acceptable degrees of self-sacrifice, competing contractual theories of ethics, and moral psychology.

Some specific sample questions are:

What is the place of altruism and unlimited love in philosophical traditions, and what empirical assumptions about human nature and the cosmos have been influential in this?

How have religious concepts of altruistic, compassionate, and unlimited love shaped later secular conceptions of ethics and of other-regarding attitudes and behaviors?

Biology

Altruistic and unlimited love are not well understood biologically. Although this love is visible and palpable (like the tip of an iceberg), we understand very little of what lies under the water line. It is time to marshal the capacities of biology and all the life sciences to better understand the embodiment of unlimited love, as well as the evolution of these capacities and the emergence of such love.

Studies pertaining to the endocrinological, neurological, immunological, genetic, and all other biological aspects of altruistic or agapic love, either as given or as received, should be encouraged, including research on the evolutionary origins of altruistic behavior, on the relationship of received love to emotional memory, and on the impact of received love on child neurological development and thriving. A specific example would be the study of parental love, the "strong" form of altruism (according to evolutionary biology), which may serve as the biological underpinning of love for those who are not genetically related to the agent.

Some specific sample questions are:

What are the evolutionary origins and neurological substrates of altruism and unlimited love? How might these interface with cultural, religious, and social factors?

What are the physiological correlates of altruistic love, both given and received?

What role does attachment, bonding, or empathy play in the expression of altruistic love?

Psychology/Human Development

Altruistic and unlimited love are experienced psychologically by the recipient in life-transforming ways that are associated with peace and well-being. Individuals who have never experienced love or compassion may be extremely hostile and abusive in their responses to the world, and eventually reach a state of crisis. Yet after experiencing even a brief period of love, they may undergo a dramatic transformation in which they migrate from hatred to *agape*. We know very little about the transforming power of love, or the inner reorientation of its recipient that turns him or her into its agent. Psychology and religion both speak of the experience of redemption through an accepting, unlimited, and unconditional form of love that does not discriminate against the recipient, no matter how sordid his or her past.

Studies on the redemptive features of altruistic and *agape* love should be encouraged. We need to understand how such love causes change in the recipient, how long the love must be sustained, how lasting this change is, and what the psychological health benefits of this transformation are with regard to sense of worth and self-esteem. We should encourage studies on how receiving such love unleashes the capacity to love, thereby producing a shift from egoism to altruism, and on how hatred, fear, anger, and resentment are reduced.

The psychological study of altruistic love also includes developmental psychology. Abuse and other forms of domestic violence sometimes rob the developing child of any opportunity to experience and learn altruistic love, care, and compassion. In the absence of such love, bonding cannot occur. The child may miss all the nonverbal expressions of love, such as through caring touch, affective tone of voice, facial expression, and the like. A child who has suffered severe peer rejection during adolescence may also become hostile and even violent. Children and young people who have not experienced or seen love expressed are unlikely to manifest it themselves. Some individuals will not manifest altruistic love until they face a severe life-threatening illness or the imminence of death in old age, if at all. Studies should be encouraged on the experiences of giving and receiving altruistic and unlimited love at all points in the human life cycle, including those aspects of each stage of the life cycle that seem to enhance or diminish growth in love.

Some specific sample questions are:

What developmental processes foster or hinder altruistic attitudes and behavior in various stages of life from early childhood onward? What role does emotional and social learning play in these processes?

What can cognitive neurosciences and developmental psychology contribute to our understanding of altruistic, compassionate, and unlimited love? For example, how do narrative, symbol, and different views of reality influence our capacities to respond in love to various situations, and to those who are neither kin nor friend but who are in genuine need?

How do emotions and altruistic love interrelate? Which emotions support altruistic love, and in what circumstances? Which emotions in what circumstances inhibit altruistic love?

Education

There is much debate about how and whether to teach a life of altruistic and unlimited love. Historically, this has been the province of spiritual and religious traditions, particularly in their presentation of the lives of the saints. It is not clear that altruism can be taught in the absence of such images of human fulfillment, both secular and sacred. Research on education and altruism interfaces with psychology and human development, yet constitutes a unique set of concerns.

Some specific sample questions are:

What are the perennial roles of religious traditions in teaching altruistic and agapic love, as well as any other historical means by which such pedagogy has been implemented?

Does love need to be seen and experienced before it can be learned?

Does engaging young people in social benevolence efforts tap their altruistic capacities in ways that affect their entire lives?

How deeply does a culture of violence and hostility adversely impact the emergence of altruistic behavior?

What is the role of mentoring in altruistic love, and can the study of contemporary altruists enhance the manifestation of unlimited love?

Anthropology / Sociology / Political Science / Economics

Social scientists have long tried to prove or disprove the very existence of motivational altruism through elaborate studies of human behavior in circumstances in which another person, often a perfect stranger, is in dire need of help. Anthropologists have compared more altruistic cultures with less altruistic ones. Economists have, since Adam Smith, attempted to analyze and balance the social sentiments and self-interested rational choice.

Studies should be encouraged on the extent to which human individuals and societies manifest behavior that is motivationally or consequentially altruistic; on what social and cultural factors influence the emergence of altruism, or counteract it; on how much, if at all, altruists are limited by in-group tendencies that may give rise to hostility toward out-groups; on how altruistic love expressed in the spheres of family and friends can be expanded to include all humanity; on what can be learned from cultures in which a remarkable amount of altruistic and unlimited love is manifest; on how can we better understand the link between love and its manifestation in compassion, care, and service.

Some specific sample questions are:

In what ways might proper self-love and neighbor-love reinforce one another?

How does altruistic love interact with pro-social motivations?

What means are available to expand or extend altruism and unlimited love to those thought of as outside one's social group? How do we define the "outsider" and how does this influence our attitudes and actions?

What role does the media play in encouraging or discouraging altruism and altruistic love?

How do particular visions of reality and worldviews affect attitudes and expressions of altruism and unlimited love?

What can economic research tell us about the nature and expression of altruism and altruistic love? How does such love affect our attitudes and behaviors toward money and the use of wealth? What is the basis of philanthropy and can it be successfully encouraged?

How might models of human action (for example, utility maximization and profit maximization in economics, "game theory" in social science and biology) as well as cultural assumptions affect the expression of altruism and altruistic love in human relations and social structures?

Medical

Altruistic and unlimited love have long been associated with helping in the recovery from various forms of physical and mental illness. Studies should be encouraged on the physiological health impact of altruistic and agape love that is given or received.

Some specific sample questions are:

How does the giving or receiving of altruistic, compassionate, and unlimited love affect mortality?

How does such love affect persons with mental or physical illnesses, especially in severe cases?

How does the receiving of such love affect persons with cognitive deficits—for example, those with retardation or dementia, or with serious mental disorders?

To what extent are health care professionals motivated by altruistic love, and how does this affect them and their patients?

How do altruism and altruistic love enter into the context of organ donation, in which the donation of organs is viewed as a "gift of life" for the stranger in need?

Conclusion

A summary of projects thus far funded by the Institute for Research on Unlimited Love is provided in chapter 11. These projects represent only a small subset of the topics that are of significance in this field. All funded research contributes, directly or indirectly, to the emerging dialogue between the science, practice, and metaphysics of Unlimited Love. The time has arrived for the study of *agape* love to move forward into the interface with science if this ideal is to gain the credibility that it merits. This book will hopefully contribute to such integration.

11

FUNDED RESEARCH PROJECTS:
2003–2005

T HE INSTITUTE for Research on Unlimited Love (www.unlimitedlove
institute.org) has taken its first concrete steps toward contributing to
scientific research on the human capacity for unselfish love by funding 21
scientific projects. These 21 projects were selected from a group of 85 full
applications, which had been invited from the more than 320 Letters of
Intent received in March 2002 in response to a nationally disseminated
Request for Proposals. After a painstaking review of each proposal by two
national experts and the Institute's research area consultants, $1,730,000
was awarded. These awards represent dramatic and steady progress for
the not-for-profit Institute, begun in 2001 with an initial endowment of $4
million from the John Templeton Foundation. The Institute is located at
the School of Medicine at Case Western Reserve University in Cleveland,
and has established itself as a center for scientific research on unselfish
love for all humanity.

The Institute offers the following definition of unlimited love: *The essence
of love is to affectively affirm as well as to unselfishly delight in the well-being of others,
and to engage in acts of care and service on their behalf; unlimited love extends this love
to all others without exception, in an enduring and constant way. Widely considered the
highest form of virtue, unlimited love is often deemed a Creative Presence underlying and
integral to all of reality: participation in unlimited love constitutes the fullest experience
of spirituality.*

At a time when greed, hatred, and group violence might lead us to doubt
its potential, we must rededicate ourselves to progress in unlimited love
and its manifestation in the world. We can be heartened by the many average people who routinely act in remarkably compassionate ways, and

respond nobly in response to the hatred manifested in events such as September 11th or the Columbine shootings. We can be inspired by all those who live lives of exceptional loving-kindness.

FUNDED PROJECTS

Area One: Human Development

Four research studies have been funded in the area of other-regarding love and human development. These include: a psychological study of autism aimed at illuminating the process of affiliation as a precursor to love; a psychological study of the mother-child relationship and its effects on the human development of empathy, with special attention to the variable of maternal spirituality; a sociological study of adolescents investigating relational and environmental contexts (including the effects of spiritual resources) and their impact on the development of other-regarding love in this age group; a psychological study examining the potential benefits of spiritual attachments and altruistic behaviors in the general adult population traumatized by the violent tragedy of 9/11/01.

Together, these studies will contribute to a better understanding of the development of unselfish love and its relationship to psychological well-being.

Research Area Consultant
Gregory Fricchione, M.D., IRUL's research area consultant for human development, is a psychiatrist specializing in medical and neuropsychiatry. He is Associate Professor of Psychiatry at Harvard Medical School and Associate Chief of Psychiatry at Massachusetts General Hospital in Boston. He is also Director of the Division of Psychiatry and Medicine at Massachusetts General Hospital. His basic research interests have centered on immune cell behavior and nitric oxide effects. His clinical research has focused on the catatonic syndrome and on the interface of psychiatry and medicine in cardiac and other diseases, where attachment behavior in the doctor-patient relationship is of key importance. In the last several years, he has been working on a project that examines the connection between brain evolution and the human spiritual imperative. Before returning to Boston in 2002, he spent two years as Director of the Carter

Center Mental Health Program in Atlanta, working on domestic and international public mental health projects.

Gregory L. Fricchione, M.D.
Director, Division of Psychiatry in Medicine
Department of Psychiatry, Massachusetts General Hospital
55 Fruit Street
Boston, MA 02114
Tel. 617-724-7816
Email gfricchione@partners.org

1. The Comprehension of Love and Altruism in Autistic and Normal Children
Jerome Kagan proposes a powerful, creative approach to peeling away the mystery of autism by determining whether the empathic deficits of the child with autism lie in the mind, behavior, or both. Much can be learned about the human development of empathy, altruism, and love from the psychophysiological and behavioral study of those with dysfunctions in these attributes. This project will study children with apparent dysfunction in social affiliation due to autism and those on the normal spectrum in this regard.

Jerome Kagan, Ph.D.
Psychology Department
Harvard University
William James Hall
33 Kirkland Street
Cambridge, MA 02138-2044
Tel. 617-495-3870
Email jk@wjh.harvard.edu

2. Love, Emotion and Empathy: Infancy to Early Childhood
Alan Fogel will examine how the quality of the mother-child relationship at 1, 2, 3, and 5 years is related to empathy development at age 5. The capacity to be attuned to others' emotions begins in relationships marked by secure attachment and co-regulation of feelings. The empathic capacity presupposes the ability to become part of a larger whole. It may be encouraged in families in which spirituality and religion play a larger role. Fogel

hypothesizes that empathy at age 5 will correlate with a high stable or a rising pattern of co-regulation and secure attachment between ages 1 and 5. Spiritual and religious well-being and engagement will be entered into the analysis to check for moderating effects. This project will provide much-needed insight into the development of empathy in children, as well as the importance of mother-child attunement and maternal spirituality.

Alan Fogel, Ph.D.
Department of Psychology
University of Utah
390 S. 1530 E. (room 502)
Salt Lake City, UT 84112-0251
Tel. 801-581-8560
Email alan.fogel@psych.utah.edu

3. Cultivating Adolescents' Other-Regarding Virtues: The Developmental Pathways to Unlimited Love

This study by Peter Benson seeks to understand the linkages among the ecologies of youth that promote, discourage, or remain silent on altruistic love, other-regarding virtues, and actions that are designed to enhance the welfare of others. The bioecological systems model of Bronfenbrenner forms its theoretical base. Two existing data sets will be used: a cross-sectional data set of 229,000 adolescents, and a longitudinal set of almost 400 adolescents assessed at three points in time. Both data sets contain responses to the Profile of Student Life: Attitudes and Behaviors (PSL-AB), which was designed by the Search Institute to assess developmental assets. This study examines the developmental ecologies of families, religious institutions, schools, neighborhoods, local communities, and non-parental adults with regard to other-regarding dispositions and helping behaviors. Spiritual and religious assets are important variables to study.

Peter L. Benson, Ph.D.
President, Search Institute—Practical Research
    Benefiting Children and Youth
The Banks Building
615 First Avenue N.E., Suite 125

Minneapolis MN 55413
Tel. 612-399-0223
Email peterb@search-institute.org

4. What Love Has to Do with It: Altruism, Generativity and Spirituality in the Aftermath of 9/11/01

The principal investigators of this study will study altruism, generativity, and spirituality in a sample of 3,000 respondents to a web-based questionnaire. Quantitative measures include: the 9/11 specific coping questionnaire; Brief COPE; Posttraumatic Growth Inventory; Multidimensional Measure of Religiousness and Spirituality; Scale of Psychological Well-Being; Social Well-Being Scale; Brief Symptom Inventory; PTSD Checklist; demographics; exposure extent. Qualitative data for 100 subjects will be gathered for linguistic analysis. The researchers hypothesize that those with higher altruism and generativity, and those who draw upon more spiritual resources at the outset, will have less posttraumatic stress disorder (PTSD) and less general symptom distress at baseline and at follow-up. This study will be a part of a Stanford University 9/11 project being led by David Spiegel, M.D. Spirituality will be assessed in terms of global religiousness/spirituality, religious coping, and spiritual change. This project hypothesizes a connection between aspects of other-regarding love and human resilience in the face of trauma and tragedy.

Cheryl Koopman, Ph.D.
Department of Psychiatry and Behavioral Sciences
Stanford University School of Medicine
401 Quarry Road, Rm. 2327
Stanford, CA 94305-5718
Tel. 650-723-9081
Email koopman@stanford.edu

Lisa D. Butler, Ph.D.
Department of Psychiatry and Behavioral Sciences
Stanford University School of Medicine
401 Quarry Road, Rm. 2327
Stanford, CA 94305-5718
Tel. 650-498-5528
Email butler@psych.stanford.edu

Area Two: Public Health and Medicine

Three research studies have been funded in the public health and medicine section. These include: a clinical intervention study in breast cancer patients and their partners; an epidemiologic case-control study of military veterans experiencing PTSD; a sociological study of how broken lives are healed and empowered among participants in a charismatic church ministry program. Together, these investigations will document how love impacts on physical and psychological well-being across the natural history of disease in both clinical and community settings. This work promises to start a new field of medical research concerned with the health effects of love.

Research Area Consultant

Jeff Levin, Ph.D., M.P.H., an epidemiologist and former medical school professor, is IRUL's research area consultant for public health and medicine. Beginning in the 1980s, his research helped create the field of religion, spirituality, and health. He is the author of over one hundred thirty scholarly publications, as well as the popular book *God, Faith, and Health*. Dr. Levin is currently researching historical and theological perspectives on what it means to love and be loved by God, as well as the physical and mental health effects of such a loving relationship.

Jeff Levin, Ph.D., M.P.H.
13520 Kiowa Road
Valley Falls, Kansas 66088
Tel. 785-945-6139
Email Levin@grasshoppernet.com

1. Effects of Compassionate/Loving Intention as a Therapeutic Intervention by Partners of Breast Cancer Patients: A Randomized Controlled Trial

Ellen G. Levine, Ph.D., M.P.H., medical psychologist at California Pacific Medical Center, is principal investigator of this research project. This study will investigate the effects of compassionate loving intention by partners of breast cancer patients on a variety of health and healthcare outcomes; it will also examine quality-of-life indicators in both patients and partners. The study will include measures of functional health, medical services utilization, psychological and spiritual well-being, marital satisfaction, physiological response to stress, and several psychological tests.

Stage I or II breast cancer patients and their partners will be recruited from the San Francisco Bay Area and randomized into experimental and control groups. Experimental-group partners will be given a training workshop structured to enhance their ability to provide loving compassion. Supported by daily home practice for three months, the training will consist of guided instruction in several meditative and mental focusing approaches, including a Tibetan Buddhist breath-based technique for eliciting compassion and LeShan type I healer training.

This project will provide an excellent opportunity to examine whether an intervention designed to strengthen the sense of self-efficacy in partners of breast cancer patients can enhance their success as caregivers. If systematic training in techniques of loving compassion is shown to be effective, it may offer a means of improving the care of people suffering from a wide range of chronic illnesses.

Ellen G. Levine, Ph.D., M.P.H.
California Pacific Medical Center
2300 California St., Suite 204
San Francisco, CA 94115
Tel. 415-600-1447
Email elevine@cooper.cpmc.org

2. Care for the Soul: The Role of Divine Love and Human Love in Adjustment to Military Trauma

Robert Hierholzer, M.D., psychiatrist with the Veterans Affairs Central California Health Care System, is principal investigator of this project, which is a longitudinal epidemiologic investigation of the protective effects of divine and human love on adjustment to military trauma among U.S. veterans.

Study subjects will be recruited from outpatient veterans at VA clinics in the Fresno area. A total of 100 case subjects who meet DSM-IV criteria for military-related posttraumatic stress disorder (PTSD) and 100 control subjects who do not meet these criteria will be sampled from this population. Participants will be given a battery of health-related tests—i.e., assessments related to PTSD, psychopathology, and symptomatology, as well as numerous psychosocial scales. These will include a set of validated measures assessing the presence of loving relationships with God and other people. Using a case-control design and epidemiologic methods of analysis, investigators will explore the relationships among different types of loving attachments, level of combat exposure, and development of current military-related PTSD in veterans.

This project promises to make exciting contributions to clinical care for sick veterans and to the validation of theoretical work in psychology that proposes salutary effects for secure attachments to significant others. Additionally, results should advance our understanding of the etiology and prognosis of PTSD.

Robert Hierholzer, M.D.
Chief, Mental Health
VACCHCS-Fresno
Department of Psychiatry
(116A) 2615 E. Clinton
Fresno, CA 93703
Tel. 559-225-6100 x5532
Email Robert.Hierholzer@med.va.gov

3. Charismatic Empowerment and Unlimited Love: A Social Psychological Assessment

Margaret M. Poloma, Ph.D., Emeritus Professor of Sociology at the University of Akron, is principal investigator of the project, which entails a multifaceted investigation of dimensions of love, religious experience, and mysticism within a charismatic Christian church community serving Atlanta's poor.

Using a variety of qualitative and quantitative methodological approaches, Dr. Poloma will conduct a longitudinal evaluation of the church's training program, which seeks to rebuild and heal broken lives by empowering people with spiritual gifts. A centerpiece of this study will be the psychometric development and validation of a new multidimensional scale of love that is based on the work of sociologist Pitirim Sorokin and others. A battery of questions will be given to at least 200 respondents; the resulting scale will be used in subsequent analyses.

Results of this study will make an important contribution to research in the sociology and psychology of religion, as well as to ministries seeking to reach out to disadvantaged individuals through religiously-grounded loving.

Margaret M. Poloma, Ph.D.
Sociology Department
The University of Akron
Akron, Ohio 44325-1905
Tel. 330-923-7860
Email mpoloma@uakron.edu

Area Three: Mechanisms by Which Altruistic Love Affects Health

The chain of events leading from negative emotional responses to external events to impact on disease has been well investigated. There has been less research into applying the same model to studying the effects of positive emotional responses on health. The same approaches and standards that have been used in the stress literature can be applied to the study of the chain of events leading from altruistic love to beneficial effects on health. It is possible that the mechanism by which altruistic love affects health occurs through blocking or attenuating the stress response, or through activating positive neurotransmitter pathways in the brain.

Research Area Consultant

Esther M. Sternberg, M.D., is our IRUL research area consultant in this area. She was trained at McGill University and practiced medicine in Montreal. She then returned to a research career and teaching at the Washington University School of Medicine in St. Louis. Her recent book, *The Balance Within: The Science Connecting Health and Emotions*, has been universally well received as one of the best books on emotions and health.

Esther M. Sternberg, M.D.
Professor, American University

1. The Physiology of Love: Empathic Responding to Emotional Reactions

Stephanie D. Preston's highly original and exciting project studies empathy from a perception-action perspective. That is, it postulates that empathy is a biological process that involves a set of specialized nerve cells in the brain that allow one to mimic motor actions, emotions, and social behaviors. These nerve cells, called mirror neurons, have been well studied in the context of perception-motor response—i.e., the phenomenon that allows a person to watch and mimic the actions of others, much as in the child's game "Simon Says." This study proposes that a similar process, utilizing the same sorts of neurons, may underlie the biological process of empathy.

The grant proposes to use a storytelling situation, combined with neuro-imaging, psychological instruments, and objective physiological measures, to compare subjects selected from different professions with high empathic components (firemen, ministers, and doctors) with others, and with a group of brain-damaged patients.

Stephanie D. Preston, Ph.D.
University of Iowa Hospitals and Clinics
200 Hawkins Drive
2 RCP—Neurology Department
Iowa City, IA 52242
Tel. 319-384-5934
Email stephanie-d-preston@uiowa.edu

2. Toward an Understanding of the Neurobiology of Parental Love
This project, conducted by James F. Leckman, M.D., proposes to compare some aspects of the neurobiology of parental love with the processes involved in obsessive behaviors. It uses a very powerful naturalistic situation to study these interactions—i.e., an infant's cry and the visual stimulus of seeing the infant. This project addresses parental love behaviors and their neural and neuroendocrine underpinnings and postulates that these may be biologically set to focus and perpetuate a connection between the parent and child. The study will use neuro-imaging (MRI), psychological instruments, and physiological hormone measures (oxytocin and cortisol) known to be activated in association with such behaviors. The project is likely to yield important information regarding the neurobiology of pathways of love, and those elements of love that resemble the more extreme behaviors that can be seen in obsessive-compulsive disorder (OCD). The theory that extreme behaviors seen in OCD may in part stem from dys-regulated biological pathways that evolutionarily evolved to cement parent-infant relationships is novel and exciting; if validated, it will serve to change

our thinking about both the state of love and OCD. It could in fact contribute to a paradigm shift in the field by relating some aspects of parental love and empathy with some obsessive behaviors. When parental love and empathy are appropriately applied in the measured amounts and contexts, they are necessary and adaptive for both parent and offspring; when inappropriate or excessive, they may constitute disease.

James F. Leckman, M.D.
Director of Research
Neison Harris Professor of Child Psychiatry,
  Pediatrics and Psychology
Yale University Child Study Center
230 South Frontage Road
P.O. Box 207900
New Haven, CT 06520-7900
Tel. 203-785-7971
Email james.leckman@yale.edu

3. Is There a Neurobiology of Love?

This project, proposed by Dr. Sue Carter, Psychiatric Institute, Chicago, is an extremely well-designed animal study in a model that has shown that the hormones oxytocin and vasopressin play an important role in affiliative behavior and development of social bonds between parent and offspring. While it is difficult to devise ways to study empathy and love in animals, this project provides an extremely well-controlled approach to understanding the precise relationships between different brain hormones and social bonding. This model will definitely shed light on the role of oxytocin in these behaviors. Most importantly, the project will examine the health benefits of loving interactions, which have been observed in epidemiological studies in humans but are difficult to address in a systematic way in human studies. Preliminary data indicate that female animals exposed to pups show a lower level of the stress hormone cortisol. Oxytocin, one of the other hormones that will be studied, may mediate these anti-stress effects. The experimental approach is novel in that it measures a mother's hormonal and behavioral responses to a naturalistic setting—i.e., exposure to a pup—to determine the health effects on the mother. This project also considers the extent to which generalized love for humanity has a hormonal basis.

C. Sue Carter, Ph.D.
Professor of Psychiatry and Co-Director
The Brain-Body Center
Department of Psychiatry
The Psychiatric Institute (MC 912)
1601 West Taylor Street
Chicago, IL 60612
Tel. 312-355-1593
Email scarter@psych.uic.edu

Area Four: Other-Regarding Virtues

Recent advances in the social sciences point to character traits or dispositions that equip people for success in the interpersonal world. These traits, which some philosophers have called "virtues," include trust, generosity, faith, empathy, kindness, gratitude, forgiveness, and honesty, among others. Such traits are presumed to help people live lives in which they are useful to other people, seek just solutions to social dilemmas, and care for the welfare of others. These other-regarding virtues may also foster physical health or psychological and relational well-being. Collectively, funded projects in the other-regarding virtues area shed light on how such virtues can be facilitated in laboratory and applied settings, and how they influence physical health, psychological well-being, and interpersonal relations. The projects are distinct from much of the "mainstream" social-scientific work on these topics in that they explore distinctively religious or spiritual contours of other-regarding virtues under investigation.

Research Area Consultant
Michael E. McCullough, Ph.D., is the IRUL research consultant in this area. He is an Associate Professor of Psychology, with a secondary appointment in the Department of Religious Studies, at the University of Miami, Coral Gables, Florida. He has authored over sixty scientific articles and book chapters on religion, spirituality, and the virtues, including work on the relationships of such variables to physical health, psychological well-being, and interpersonal relations. He has also written and edited several books on these subjects.

Michael E. McCullough, Ph.D.
Department of Psychology
University of Miami
P.O. Box 248185
5202 University Drive
Coral Gables, FL 33124-2070
Tel. 305-284-8057
Email mikem@umiami.edu

1. Other-Regarding Love for Individuals Outside One's Social Group
Stephen Wright and Arthur Aron will conduct a study at the University of California, Santa Cruz, and at the State University of New York at Stony Brook that examines love for people outside one's own social group, the absence of which is one of the world's most grievous and seemingly intractable moral and spiritual problems. Hope for addressing this problem in novel and effective ways comes from a psychological model of close relationships that was originally inspired by the Upanishads and has recently been applied to intergroup relations. The central idea of the model is that close others and those in one's social groups function in a sense as part of oneself; the regard and caring that one usually experiences for oneself is thereby extended to close others. Further, the social identities of close others become to some extent one's own. As a result, one becomes more inclined to extend caring and love to the friend's ethnic group. This project will explore this phenomenon by examining (a) variables such as caring, empathy, and trust toward outgroup members; (b) the specific role of inclusion of other in the self as the mechanism underlying the effects of cross-group friendship on prejudice toward members of that outgroup; (c) the possible moderating role of religiousness/spirituality; (d) the practical potential for applying these concepts in the real world. Three studies will be conducted: a laboratory study that creates interpersonal closeness between people of different ethnic groups; a survey of students' friendships with people from different ethnic groups; an applied study designed to increase students' other-regarding love for members of other ethnic groups by using established laboratory procedures for fostering inclusion of the others in the self.

Stephen C. Wright, Ph.D.
Psychology Department
Social Sciences II
University of California, Santa Cruz
Santa Cruz, CA 95064
Tel. 831-459-3557
Email swright@cats.ucsc.edu

Arthur Aron, Ph.D.
Department of Psychology
State University of New York at Stony Brook
Stony Brook, NY 11794-2500
Tel. 631-632-7707
Email Arthur.Aron@sunysb.edu

2. The Gift of One's Self: Expressions of Unlimited Love and Gratitude in Organ Donors and Recipients

Robert A. Emmons will study organ donation, often referred to as the "gift of life." The overall goal of this project is to examine expressions of unlimited love in the form of organ donation and the role that the virtue of gratitude plays in motivating donation and recipient behavior. The specific aims of the project are to: (a) investigate the degree to which self-transcendent strivings (spirituality, intimacy, and generativity) predict intentions to donate organs and actual organ donation; (b) test the "moral motive" hypothesis of gratitude: Does the virtue of gratitude for life predict intention or willingness to donate part of one's self?; (c) examine whether an intervention designed to increase gratitude increases actual intention to become an organ donor; (d) examine whether the expression of gratitude by transplant recipients increases their likelihood of thriving post-transplant. The project is strengthened by the diversity of its methods. The investigators will incorporate correlational, prospective longitudinal, qualitative, and experimental methods.

Robert A. Emmons, Ph.D.
University of California, Davis
Department of Psychology

One Shields Avenue
Davis, CA 95616-8686
Tel. 530-752-8844
Email raemmons@ucdavis.edu

3. The Self as a Conduit of Love

Julie Juola Exline will examine the ways in which receiving love from others enhances one's ability to love. Love is a common thread that underlies many virtuous actions, including helping behavior, emphasis on positive qualities in others, and forgiveness and apology in the wake of offenses. This project will test a conduit model of altruistic love. The model predicts that we are most able to love if we have first received love, either from other people or from God. The proposed research also addresses the role of grace, or undeserved favor, in the transmission of love. Studies will be primarily experimental, beginning with laboratory-based designs and culminating in an intervention study. Laboratory-based studies will address whether feeling loved—especially when the love is seen as undeserved—motivates people to return love to the source. A second set of studies will address whether people who receive love, and are reminded to pass it on, will become more loving to third parties. Finally, an intervention will be developed to give participants the tools to love in situations in which doing so would be difficult. Religious themes will be emphasized, including participants' relationships with God. One major aim of the proposed project is to provide a bridge between scientific and theological literatures on the topics of altruistic love, forgiveness, justice, and grace. By focusing on the dual roles of giving and receiving love, the long-term aim of the project is to give people practical tools that will enhance their well-being, their perceived relationships with God, and their ability to love others.

Julie Juola Exline, Ph.D.
Department of Psychology
11220 Bellflower Road
Case Western Reserve University
Cleveland, Ohio 44106-1723
Tel. 216-368-8573
Email jaj20@po.cwru.edu

Area Five: Evolutionary Perspectives on Other-Regard

Evolutionary biology has a unique relationship to the issue of unlimited love for two reasons. First, from Darwin on, sacrificial behavior has been recognized as a crucial question for evolutionary theory. Many contemporary accounts have tended to dismiss altruism as an end, or even a possibility, of human existence, because this has been understood to be a core entailment of evolutionary theory. Second, over the last generation, evolutionary theory has dramatically influenced other academic disciplines, and has been turned to by popular media for authoritative exegesis of the human condition. Thus, evolutionary biology is crucially important to popular and scholarly discussions of love. Recent promising approaches to the evolutionary elucidation of altruism include multilevel selection theory, econometric and evolutionary game theory, comparative anthropology, and behavioral studies of non-human primates. IRUL is funding seminal work in each of these four areas.

Research Area Consultant

Jeffrey P. Schloss, Ph.D., serves as IRUL research consultant in this area. He received his Ph.D. in Ecology and Evolutionary Biology from Washington University, and has taught at the University of Michigan, Wheaton College, and Jaguar Creek Tropical Research Center; he is now Professor and Chair of Biology at Westmont College in Santa Barbara. He has been awarded a Danforth Fellow, an AAAS Fellow in Science Communication, and serves on the editorial and advisory boards of numerous journals and organizations relating science and religion. He is interested in evolutionary theories of human nature. His most recent projects include a collaborative volume just released from Oxford University Press, *Altruism and Altruistic Love: Science, Philosophy, and Religion in Dialogue*, and coeditorship of a two-volume *Journal of Psychology & Theology*, focusing on biological and theological perspectives on human nature.

Jeffrey P. Schloss, Ph.D.
Department of Biology, Westmont College
955 La Paz Road
Santa Barbara, CA 93108
Tel. 805-565-6118
Email schloss@westmont.edu

1. Altruistic Love, Evolution, and Individual Experience

Evolutionary theory tends to be theory rich, but data poor: in comparison, the human behavioral sciences are data abundant, but lack a unifying theoretical foundation. David Sloan Wilson will apply the theoretical perspective of multilevel selection to the interpretation of data of life experience in one of the most voluminous databases available—the experience sampling method (ESM) of Mihaly Csikszentmihalyi. ESM is to psychological life experience what integrated cross-cultural databases are to anthropological assessment. Wilson's group selectionist model posits that human groups are significant functional units that facilitate the emergence of capacities for both genuine sacrifice and defection. This allows a variety of testable predictions about the relationship among altruism, religion, life stress, and other variables. Wilson's proposal is the first attempt to test these predictions on a large scale with highly-regarded data. This study is likely to be landmark in its use of data that poll life experience and establish its relationship to altruism.

David Sloan Wilson
Department of Biological Sciences
Binghamton University (SUNY)
Binghamton, NY 13902-6000
Tel. 607-777-4393
Email dwilson@binghamton.edu

2. Unlimited Love in the Laboratory: Evaluating the Effect
of Religion on Sharing and Cooperative Behavior

Peter Richerson proposes to test the relationship among religious experiences, beliefs, and involvements in cooperative sacrifice by unifying two well-developed and never before integrated approaches of research: game theory experiments and psychometric religious assessment. In two different phases involving student subjects and members of religious and nonreligious communities, subjects will be given a variety of standardized measures of religious experience, belief, and involvement, and subjected to two classic game theoretic experiments: the Ultimatum Game (which measures cooperative fairness and altruistic punishment), and the Commons Game (which assesses commitment or detraction from the common good). These tools will be used to examine how sacrificial behavioral patterns relate to self-reported varieties of religious experience, religious belief, and religious involvements. Group selection theory suggests that increased intra-group commitment will result in more in-group sacrifice and out-group rejection. This theory will be tested with an experimental design intended to illuminate the relationship between group loyalty and expansive sympathy.

Peter Richerson, Ph.D.
Department of Environmental Studies and Policy
University of California, Davis
One Shields Avenue
Davis, CA 95616
Tel. 530-752-2781

3. Cross-Cultural Survey of Altruistic Behavior

Christopher Boehm proposes to compile, tabulate, and assess a landmark cross-cultural database of cooperative behaviors in Paleolithic-representative hunter-gatherer societies. Out of 339 available h-g cultures, he has chosen 154 that are credible as representatives of Paleolithic ancestry due to lack of contact with agricultural or industrial influences. He will scan and code ethnographies for a wide variety of kin, reciprocal, non-reciprocal in-group/out-group cooperation, plus variables relating to religion and moral social controls. This work is important because the empirical basis for sociobiological theories of human nepotism and strict reciprocity is largely untested or relies on a limited selection of available ethnographies. Boehm will assemble an exhaustive database, with extensive coding for altruism and related parameters, in order to test competing theories of the origin, nature, and maintenance of altruism. This work may provide an empirical basis for an understanding of human love.

Christopher Boehm, Ph.D.
Director, The Jane Goodall Research Center
Department of Anthropology
University of Southern California
Los Angeles, CA 90089
Tel. 213-740-1900
Email cboehm1@msn.com

4. An Evolutionary Perspective on the Emotional Prerequisites for Love

Love requires that one examine the situation of the other; while empathy may not be sufficient, it is certainly a necessary building block for other-regard. Capacities underlying altruistic love and compassion build upon a human psychological architecture that has been shaped by evolutionary history. If we wish to learn more about the evolution of constituent capacities of love, it is important to understand expressions of empathy and sympathy in other animals. The chimpanzee, our closest relative, exhibits evidence for "consolation behavior," defined as a bystander providing reassuring contact to a distressed conspecific. We don't, however, understand the underlying motivation, which could entail simple emotional contagion or extension of sympathetic concern. Behavioral predictions for these two models differ. The proposed research involves a behavioral study—with both observational and experimental components—designed to distinguish between different sources of empathic response in this closely related primate.

Frans B. M. de Waal
Director, Living Links Center
Yerkes National Primate Research Center
Emory University
954 N. Gatewood Road
Atlanta, GA 30329
Tel. 404-727-7898
Email dewaal@emory.edu

Area Six: The Sociological Study of Faith-based Communities and Their Activities in Relation to the Spiritual Ideal of Unlimited Love

This topic area examines the significance of concepts of "love for all humanity" in a sociological context, giving attention to the ways in which this spiritual ideal is implemented within faith traditions through volunteerism and service to the neediest. While religious communities can and do fall short of the ideal of "unlimited love," and sometimes even descend into in-group insularity, love for all humanity is nevertheless a key precept that often translates into personal and organizational altruistic behavior.

Research Area Consultant

Byron R. Johnson, Ph.D., is the IRUL research area consultant for this area. Based at the University of Pennsylvania, he is the Director of the Center for Research on Religion and Urban Civil Society, and a distinguished senior fellow in the Robert A. Fox Leadership Program. He is also a senior fellow in the Center for Civic Innovation at the Manhattan Institute. Before coming to the University of Pennsylvania, Johnson directed the Center for Crime and Justice Policy at Vanderbilt University, and remains a senior scholar in the Vanderbilt Institute for Public Policy Studies.

Byron R. Johnson, Ph.D.
Director, Center for Research on Religion
    and Urban Civil Society of the University of Pennsylvania
3814 Walnut Street
Philadelphia, PA 19104
Tel. 215-746-7111
Email byronj@sas.upenn.edu

1. A National Study of Altruistic and Unlimited Love

The project will allow key variables on aspects of unlimited love to be added as a new module to the General Social Survey (GSS), one of the most utilized and highly respected social science surveys in the world, located at the University of Chicago. Led by Dr. Tom Smith, Director of the General Social Survey, the project will include the introduction of a pilot module in 2003 that will be based on the best data from previous research on the subject. Based on the pilot, the new module on unlimited love questions will officially be added to the GSS in 2004. Housed within the National Opinion Research Center (NORC), the GSS will bring much needed social science attention and credibility to the topic of unlimited love as NORC holds substantial capital within the academic community and well beyond. The GSS is a very large, random, representative survey that allows social scientists to analyze national level trends and patterns. This new module will provide unprecedented opportunities for junior and senior scholars to explore the relationship between unlimited love and other socially important factors, including the roles of religion, religious practices, and beliefs. Because the GSS is so accessible, it will provide researchers with quick access to some of the best social survey data available. This project has the potential to be supercatalytic by providing future researchers with nationally representative data on unlimited love, as well as hundreds of other relevant and important social science variables. Such data will make it possible to "fast-forward" the research and scholarship in the area of altruism and unlimited love—which is perhaps the main overall objective of the Institute.

Tom W. Smith, Ph.D.
National Opinion Research Center (NORC)
1155 East 60th Street
Chicago, IL 60637
Tel. 773-256-6288
Email smitht@norcmail.uchicago.edu

2. Faith-based Service Organizations, Altruistic Caregiving, and Understandings of Love

Led by the sociologist Robert Wuthnow, this study is part of a larger community study of the social agencies and churches in the Lehigh Valley of Pennsylvania. The proposed study will focus on persons volunteering for caregiving for the poor, needy, or elderly. Wuthnow hypothesizes that among people working in nonprofit agencies, caring leads to trust, and trust in turn engenders effectiveness. Wuthnow goes on to postulate that faith-based organizations (FBOs) are more likely than non-faith-based organizations to exhibit the caring and loving attitudes that lead to trust and effectiveness. The proposed research is extremely important because it is perhaps the first to provide a strong theoretical foundation for the assumption that faith-based organizations are more effective than their secular counterparts. Wuthnow plans on conducting 120 in-depth interviews with volunteers that will yield rich data on motivations, understandings of unlimited love, beliefs about God's love, and much more. Interviews will be conducted with representatives from both faith-based and non-faith-based organizations. Importantly, Wuthnow plans to relate the attitudes and behavior of the volunteers to their own religious beliefs and practices. This is an important step in helping to understand the linkages between volunteer motivation and religious commitment, as well as between religiosity and community agencies, including churches. The sophisticated nature of the study methodology and its tight theoretical underpinnings lead us to believe that Wuthnow's study will eventually become a sociological classic. In summary, this study will shed important empirical light on the relationships among faith, spirituality, and motivations toward volunteering, trust, and the efficacy of caregiving.

Robert Wuthnow, Ph.D.
Department of Sociology/Center for the Study of Religion
Wallace Hall
Princeton University
Princeton, NJ 08544
Tel. 609-258-4742
Email wuthnow@princeton.edu

3. Antecedents and Correlates of Civic Engagement for African-American Adolescents and Their Parents

The proposed study takes advantage of recently collected longitudinal data from the University of Rochester Youth and Family Project. The research is a multi-method, multi-informant investigation of civic engagement among a sample of African-American adolescents. Judith Smetana, the principal investigator, posits that there is a relationship between adolescent love and trust for parents, racial socialization, religiosity, and how these influence adolescent involvement in their communities. Civic engagement is a topic of key interest; the issues of civic engagement among African-American adolescents and their parents is particularly important. The proposed study will provide new and much-needed knowledge about the role of religion or faith-based communities in encouraging civic participation within many black communities. We need to increase our understanding of civic engagement among minority populations and adolescents who face adversity; this project does both. This important study will advance our understanding of how spirituality, religiosity, compassionate love, and concepts of social justice in family contexts become instantiated in African-American late adolescents' involvement and service on behalf of the well-being of others.

Judith G. Smetana, Ph.D.
Department of Clinical and Social Sciences in Psychology
Meliora Hall, RC 270266
University of Rochester

Rochester, NY 14627
Tel. 585-275-4592
smetana@psych.rochester.edu

4. Self-Forgetfulness in Seeking the Lost: A Sociological Study of Relentless Love and Compassionate Service at Ground Zero

In the aftermath of September 11, 2001, the country will long remember the thousands of construction workers, firemen, police, and chaplains who poured into Lower Manhattan to conduct the rescue, recovery, and clean-up operation. They worked around the clock for days in the early weeks, then in grueling twelve-hour shifts looking for survivors and the dead. At St. Paul's Episcopal Chapel on Lower Broadway in New York, located on the precipice of Ground Zero, some five thousand volunteers fed these workers, gave them sleeping quarters, comforted them, clothed them, and built a spiritual community of mutual gratitude. What motivated these particular individuals to volunteer for this work? What human attributes were displayed in greatest abundance? With all the array of resources at Ground Zero, why did these persons make the Chapel their home? What was it about the experience of life in the Chapel that sustained the massive work? This study will provide scientifically based explanations for questions surrounding such notable and sustained altruistic behavior. Led by Dr. Courtney Cowart, a theologian in the St. Paul's ministry at Ground Zero, and Dr. Bambi Schieffelin, a cultural anthropologist and linguist, this important study will document the role that religious perceptions may have played in motivating and sustaining this remarkable human response to the tragedy of September 11.

Courtney V. Cowart, Th.D.
Center for Christian Spirituality
The General Theological Seminary
175 Ninth Avenue
New York, NY 10011
Tel. 212-794-3119
Email nine-twelve@att.net

ADDITIONAL GRANTS

In addition to the above twenty-one grants, the Institute for Research on Unlimited Love partnered with the Fetzer Institute initiative on the Science of Compassionate Love to fund four additional studies that are currently in progress:

1. Benevolent Love and Marriage
This study examines "benevolent love" (described in classical terms as the love of true friendship and in contemporary terms as unconditional love) within long-term marriage. Benevolent love exhibits virtues of temperance, fortitude, justice, prudence, and charity. The study will focus on the relationship of benevolent love to attractive love, marital quality and stability, and partners' religiousness.

Vincent Jeffries, Ph.D.
Department of Sociology
California State University, Northridge
18111 Nordhoff St.
Northridge, CA 91330
Tel. 818-761-7588
Email cvjeff@pacbell.net

2. Volunteerism, Community, and Compassionate Acts among Older Adults
This study examines the role of service, spirituality, religion, and older persons' personal identity in individuals at a religiously oriented retirement community and a comparison community. The interviews and self-reports will focus on the personal meanings of service, religion, altruistic love, and the role of each, and will examine whether the religiously oriented have a more highly integrated sense of concepts.

Allen Omoto, Ph.D.
Psychology Department
Claremont Graduate University
204 Academic Computing Building
123 East 8th Street
Claremont, CA 91711

Tel. 909-607-3716
Email allen.omoto@cgu.edu

3. The Development, Antecedents, and Psychosocial Implications of Altruism in Late Adulthood

This project studies altruistic love in terms of Eric Erikson's concept of generativity: the concern for and commitment to guiding the next generation. The data are from a longitudinal sample of Americans born in California in the 1920s (140 participants interviewed 4 times over the years, with the latest in 1997/2000). It will examine the vocabulary and the reasoning people use as they refer to generative/altruistic acts and the relationship of these acts to social background, personality characteristics, religion, health, and attitudes.

Paul Wink, Ph.D.
Department of Psychology
Wellesley College
106 Central Street
Wellesley, MA 02481
Tel. 781-283-3729
Email Pwink@wellesley.edu

Michele Dillon, Ph.D.
Department of Sociology
Yale University
P.O. Box 208265
New Haven, CT 06520-8265
Tel. 203-432-3320
Michele.dillon@yale.edu

4. The Impacts of Religious, Intellectual, and Civic Engagement on Altruistic Love and Compassionate Love as Expressed Through Charitable Behaviors

This study supports analysis from the 2000 Social Capital Community Benchmark Survey, which is intended to measure a U.S. representative sample of people's "social capital"—i.e., the wealth of connections among people thought to lead to pro-social behaviors and attitudes. This data analysis will examine connections between people's acts of giving and volunteering and their religious, intellectual, social, and civic development.

Eleanor Brown, Ph.D.
Department of Economics
Pomona College
425 N. College Avenue
Claremont, CA 91711
Tel. 909-607-2810
Email ebrown@pomona.edu

NOTES

Notes to the Preface

1. Pitirim A. Sokokin, *Altruistic Love: A Study of American Good Neighbors and Christian Saints* (Boston: Beacon Press, 1950), vi.

2. Edith Wyschogrod, *Saints and Postmodernism: Revisioning Moral Philosophy* (Chicago: University of Chicago Press, 1990).

3. Paul Connolly, A. Smith, and B. Kelly, *Too Young to Notice? The Cultural and Political Awareness of 3-6 Year Olds in Northern Ireland* (Belfast: Northern Ireland Community Relations Council, 2002).

4. James K. Rilling, David A. Gutman, Thorsten R. Zeh, et al., "A Neural Basis for Social Cooperation," *Neuron* 35 (July 18, 2002), 395–405.

Notes to the Introduction

1. See M. Scott Peck, *The Road Less Traveled* (New York: Simon & Schuster, 1978), especially 150–155, "The Risk of Confrontation."

2. See anthropologist M. F. Ashley Montagu's classic work, *The Direction of Human Development: Biological and Social Bases* (New York: Harper & Brothers, 1955). See also Robert Wright, *Nonzero: The Logic of Human Development* (New York: Vintage, 2002).

3. See Irenaus Eibl-Eibesfeldt and Frank K. Salter, eds., *Ethnic Conflict and Indoctrination: Altruism and Identity in Evolutionary Perspective* (New York: Berghahn Books, 1998); Patrick James and David Goetze, eds., *Evolutionary Theory and Ethnic Conflict* (Westport, Conn.: Praiger, 2001).

Notes to Chapter 1

1. See Teilhard de Chardin, *The Phenomenon of Man*, with an Introduction by Sir Julian Huxley (New York: Harper & Row, 1959).

2. M. K. Gandhi, *The Law of Love*, ed. Anand T. Hingorani (Bombay: Bharatiya Vidya Bhavan, 1970); see also Louis Fischer, ed., *The Essential Gandhi: An Anthology of his Writings on his Life, Work, and Ideas* (New York: Vintage Books, 1962).

3. Paramahansa Yogananda, *Where There is Light* (Los Angeles: Self-Realization Fellowship, 1988), 57, 137.

4. Reinhold Niebuhr, *The Children of Light and the Children of Darkness* (New York: Scribner's, 1944).

5. See Rufus M. Jones, *Spiritual Reformers in the 16th and 17th Centuries* (Boston: Beacon Press, 1914); Leo Tolstoy, *The Kingdom of God is Within You*, trans. C. Garnett (Lincoln: The University of Nebraska Press, 1984 [original 1894]); Lawrence Edward Carter, Sr., ed., *Walking Integrity: Benjamin Elijah Mays, Mentor to Martin Luther King, Jr.* (Macon, Ga.: Mercer University Press, 1998).

6. See Everett L. Worthington, ed., *Dimensions of Forgiveness: Psychological Research and Theological Perspectives* (Philadelphia: Templeton Foundation Press, 1998).

7. See Thomas Molnar, *Utopia: The Perennial Heresy* (New York: Sheed & Ward, 1967).

8. See Dietrich Bonhoeffer, *Letters and Papers from Prison* (New York: Macmillan, 1953).

9. See Millard Fuller, with Diane Scott, *No More Shacks: The Daring Vision of Habitat for Humanity* (Waco, Tex.: Word Books, 1986).

10. See Robert Wuthnow, *Acts of Compassion: Caring for Others and Helping Ourselves* (Princeton: Princeton University Press, 1991).

11. For a Buddhist perspective, see Richard J. Davidson and Anne Harrington, eds., *Visions of Compassion: Western Scientists and Tibetan Buddhists Examine Human Nature* (New York: Oxford University Press, 2002); for a Christian perspective, see M. Simone Roach, CSM, *Caring, The Human Mode of Being: A Blueprint for the Health Professions*, 2nd ed. (Ottawa, Canada: CHA Press, 2002).

12. See Jonathan Edwards, *Religious Affections*, ed. John E. Smith (New Haven: Yale University Press, 1959 [original 1746]).

13. William James, *The Varieties of Religious Experience* (New York: Penguin Books, 1982 [original 1902]), 486.

14. John M. Templeton, *Pure Unlimited Love: An Eternal Creative Force and Blessing Taught by All Religions* (Philadelphia: Templeton Foundation Press, 2000).

15. Kristen Renwick Monroe, *The Heart of Altruism: Perceptions of a Common Humanity* (Princeton: Princeton University Press, 1996).

16. Pitirim A. Sorokin, *Altruistic Love: A Study of American Good Neighbors and Christian Saints* (Boston: Beacon Press, 1950).

17. Templeton, *Pure Unlimited Love*, 3.

18. H. Richard Niebuhr, *The Purpose of the Church and Its Ministry* (New York: Harper & Row, 1977), 35.

19. Gene Outka, *Agape: An Ethical Analysis* (New Haven: Yale University Press, 1972).

20. Patty Anglin, with Joe Musser, *Acres of Hope: The Miraculous Story of One Family's Gift of Love to Children without Hope* (Uhrichsville, Ohio: Promise Press, 1999).

21. Christina Noble, with Robert Coram, *Bridge Across My Sorrows* (London: Corgi Books, 1995), 3.

22. Ibid., 17.

23. Ibid., 21–22.

24. John Woolman, *The Journal of John Woolman*, in *The Harvard Classics: Franklin, Woolman, Penn,* ed. Charles W. Elliot (New York: P. F. Collier & Son, 1937), 174.

25. Robert E. Quinn, *Deep Change: Discovering the Leader Within* (San Francisco: Jossey-Bass, 1996), 218.

26. See Friedrich Nietzsche, *On the Geneology of Morals,* trans. Walter Kaufmann (New York: Vintage Books, 1989 [original 1887]).

Notes to Chapter 2

1. See my introduction to Pitirim A. Sorokin, *The Ways and Power of Love: Types, Factors, and Techniques of Moral Transformation* (Philadelphia: Templeton Foundation Press, 2002 [original 1954]).

2. Pitirim A. Sorokin, *A Long Journey* (New Haven: College & University Press, 1963), 271.

3. Ibid., 273.

4. Ibid., 277.

5. Pitirim A. Sorokin, "Studies of the Harvard Research Center in Creative Altruism," 1955, 1–2.

6. Pitirim A. Sorokin, *Altruistic Love: A Study of American Good Neighbors and Christian Saints* (Boston: Beacon Press, 1950), 4.

7. Robert G. Hazo, *The Idea of Love*, Concepts in Western Thought Series, ed. Mortimer J. Adler (New York: Frederick A. Praeger, 1967), 286.

8. Peter Kropotkin, *Mutual Aid: A Factor in Evolution* (Montreal: Black Rose Books, 1989 [original 1903]).

9. Vladimir Solovyov, *The Meaning of Love*, ed. and trans. Thomas R. Meyer with an introduction by Owen Barfield (New York: Lindisfarne Press, 1985 [original 1894]), 43.

10. Ibid., 51.

11. Sorokin, *Ways and Power of Love*, 6-13.

12. Ibid., 14.

13. Joseph B. Ford, "Sorokin as Philosopher," in Phillip J. Allen, ed., *Pitirim A. Sorokin in Review* (Durham, N.C.: Duke University Press, 1963), 39–66.

14. Arnold J. Toynbee, "Sorokin's Philosophy of History," in ibid., 67–94.

15. See K. M. Munshi, "How Munshi of India Assesses Sorokin," in ibid., 300–317.

16. Sorokin's key texts in basic sociology include *Contemporary Sociological Theories* (New York: Harper & Row, 1928); *Social and Cultural Dynamics*, 4 vol. (New York: American Book Company, 1937–1941); *Crisis of Our Age* (New York: E. P. Dutton & Co., 1941); *Society, Culture and Personality* (New York: Harper & Brothers, 1947); and *Reconstruction of Humanity* (Boston: Beacon Press, 1948). Many of these texts were translated into dozens of languages. As he later turned to the topic of altruistic love, he published, in addition to *The Ways and Power of Love*, his major study entitled *Altruistic Love: A Study of American Good Neighbors and Christian Saints*.

17. The quotations in this section are from Sorokin, *Ways and Power of Love*, ch. 2, 16–35.

18. Ibid., 26.

19. Ibid., 84.

20. Ibid., 96.

21. Ibid., 125 (italics in original).

22. Ibid., 127.

23. Ibid., 459 (italics in original).

24. Ibid., 461.

25. Ibid., 477.

26. Ibid., 485. See Barry V. Johnston, *Pitirim A. Sorokin: An Intellectual Biography* (Lawrence: University of Kansas Press, 1996), ix.

Notes to Chapter 3

1. Leo Tolstoy, *The Death of Ivan Ilych and Other Stories* (New York: Penguin, 1960 [original 1886]), 95–156.

2. Kirk Byron Jones, *Rest in the Storm: Self-Care Strategies for Clergy and Other Caregivers* (Valley Forge, Pa.: Judson Press, 2001), 35.

3. The Dalai Lama, *Ethics for the New Millennium* (New York: Riverhead Books, 1999), 22 [italics mine], 23.

4. See Paul Ramsey, *Basic Christian Ethics* (Louisville, Ky.: John Knox Press, 1993 [original 1953]), 101.

5. Jules Toner, *The Experience of Love* (Washington, D. C.: Corpus Books, 1968), 65.

6. Ibid., 67.

7. Ibid., 79–80.

8. Ibid., 83.

9. Ibid., 104.

10. Ibid., 109.

11. Anne Harrington, "A Science of Compassion or a Compassionate Science? What Do We Expect from a Cross-Cultural Dialogue with Buddhism?" in Richard J. Davidson and Anne Harrington, eds., *Visions of Compassion: Western Science and Tibetan Buddhists Examine Human Nature* (New York: Oxford University Press, 2002), 21; see 18–30.

12. Miguel de Unamuno, *The Tragic Sense of Life*, trans. J. E. Crawford (New York: Dover, 1954 [original 1912]), 136.

13. Ibid., 139.

14. Johan Huizinga, *Homo Ludens: A Study of the Play Element in Culture* (Boston: Beacon Press, 1955).

15. William A. Sadler, *Existence and Love: A New Approach to Existential Phenomenology* (New York: Scribners, 1969), 218.

16. Ibid., 219.

17. Max Scheler, *The Nature of Sympathy*, trans. Peter Heath (London: Routledge & Kegan Paul, 1954 [original 1913], 135, 138, 152, 162.

18. Jean Vanier, *Becoming Human* (Mahwah, N. J.: Paulist Press, 1998), 11.

19. Ibid., 21.

20. Ibid., 22.

21. Ibid., 23.

22. Ibid., 26.

23. Ibid., 27.

24. Ibid., 28.

25. Tom Kitwood, *Dementia Reconsidered: The Person Comes First* (Buckingham: Open University Press, 1997), 81.

26. Ibid., 81.

27. Ibid., 83.

28. Stephen G. Post, *The Moral Challenge of Alzheimer Disease: Ethical Issues from Diagnosis to Dying* (Baltimore: Johns Hopkins University Press, 2000).

Notes to Chapter 4

1. W. E. H. Lecky, *History of European Morals from Augustus to Charlemagne* (New York: George Braziller, 1955 [original 1869]).

2. Auguste Comte, *System of Positive Polity*, vol. 1 (London: Longmans, Green & Co., 1975 [original 1851]), 556.

3. For a spiritual-theological statement on compassion, see Matthew Fox, *A Spirituality Named Compassion: Uniting Mystical Awareness with Social Justice* (Rochester, Vt.: Inner Traditions, 1979).

4. Michael Ruse, *Sociobiology: Sense and Nonsense* (Dordrecht: D. Reidel, 1979), 198.

5. Judith A. Howard and Jane Allyn Piliavin, "Altruism," in Edgar F. Borgatta and Rhonda J. V. Montgomery, eds., *Encyclopedia of Sociology*, 2nd ed., vol. 1 (New York: Macmillan Reference, 2000), 114.

6. Ibid., 114.

7. Erving Goffman, *The Presentation of the Self in Everyday Life* (Woodstock, N.Y.: The Overlook Press, 1973), 17.

8. Alisdair MacIntyre, *A Short History of Ethics* (New York: Touchstone, 1966), 135.

9. Bernard Mandeville, *The Fable of the Bees*, ed. P. Harth (Middlesex, England: Penguin Books, 1970 [original 1705]).

10. See Thomas Nagel, *The Possibility of Altruism* (Princeton: Princeton University Press, 1970).

11. Thomas E. Hill, "Beneficence and Self-Love: A Kantian Perspective," in Ellen Frankel Paul, Fred D. Miller, and Jeffrey Paul, eds., *Altruism* (Cambridge: Cambridge University Press, 1993), 1–23.

12. See L. A. Selby-Bigge, ed., "Introduction," *British Moralists: Being Selections from Writers Principally of the Eighteenth Century* (New York: Dover, 1965 [original 1897]), xi–xx, xxiii.

13. C. Daniel Batson, *The Altruism Question: Toward a Social Psychological Answer* (Hillsdale, N. J.: Lawrence Erlbaum Associates, 1991), 6.

14. Ibid., 7.

15. Ibid., 1–2.

16. John Rawls, *A Theory of Justice* (Cambridge: Harvard University Press, 1971), 13.

17. Batson, *Altruism Question*, 4.

18. Ibid., 43.

19. Ibid., 75.

20. Ibid., 81.

21. Ibid., 85.

22. Ibid., 87.

23. Ibid., 127-148.

24. Ibid., 174.

25. Charles Darwin, *The Descent of Man* (Amherst, N.Y.: Prometheus Books, 1998 [original 1874]), 100.

26. Kristen Renwick Monroe, *The Heart of Altruism: Perceptions of a Common Humanity* (Princeton: Princeton University Press, 1996), 203.

27. Ibid., 205.

28. Ibid., 9.

29. Batson, *Altruism Question*, 86.

30. Martha C. Nussbaum, *Upheavals of Thought: The Intelligence of Emotions* (New York: Cambridge University Press, 2001), 300–302.

31. Ibid., 333.

32. Robin Gill, *Churchgoing and Christian Ethics* (Cambridge: Cambridge University Press, 1999).

33. L. D. Nelson and R. R. Dynes, "The Impact of Devotionalism and Attendance on Ordinary and Emergency Helping Behavior," *Journal for the Scientific Study of Religion*, vol. 15, no. 1 (1976): 47–59.

34. B. Hunsberger and E. Platonow, "Religion and Helping Charitable Causes," *Journal of Psychology*, vol. 120, no. 6 (1986): 517–528.

35. F. M. Bernt, "Being Religious and Being Altruistic: A Study of College Service Volunteers," *Personality and Individual Differences*, vol. 10, no. 6 (1989): 663–669.

36. K. Mills, "Research and Creative Activity," *Indiana University*, vol. 22, no. 3 (January 2000).

37. Jonathan Edwards, *Treatise on the Religious Affections*, ed. John E. Smith (New Haven: Yale University Press, 1959 [original 1746]).

38. William James, *The Varieties of Religious Experience* (New York: Penguin Books, 1982 [original 1902]).

39. James W. Fowler, *Stages of Faith: The Psychology of Human Development and the Quest for Meaning* (San Francisco: HarperCollins 1981), 28.

40. See Patrick N. Juslin and John A. Sloboda, *Music and Emotion: Theory and Research* (New York: Oxford University Press, 2001).

Notes to Chapter 5

1. Richard Dawkins, *The Selfish Gene* (Oxford: Oxford University Press, 1989), 2.

2. See Nancey Murphy and George F. R. Ellis, *On the Moral Nature of the Universe: Theology, Cosmology, and Ethics* (Minneapolis: Fortress Press, 1996).

3. Catherine A. Salmon, "Kin Terms in Political Rhetoric: The Evocative Nature of Kin Terminology in Political Rhetoric," *Politics and the Life Sciences*, vol. 17, no. 1 (1998): 51.

4. George C. Williams, *Adaptation and Natural Selection* (Princeton: Princeton University Press, 1966), 95.

5. Ibid., 212.

6. Robert Axelrod, *The Evolution of Cooperation* (New York: Basic Books, 1984), 3.

7. Ibid., chapter written with William D. Hamilton, 89, 96.

8. R. C. Lewontin, "Evolution and the Theory of Games," *Journal of Theoretical Biology*, vol. 1 (1961): 382–403.

9. See Oskar Morgenstern, "Game Theory," in *The Dictionary of the History of Ideas*, vol. 2 (New York: Charles Scribner's Sons, 1973), 263–275.

10. See Gary Becker, *A Treatise on the Family* (Chicago: The University of Chicago Press, 1991).

11. Martin Nowak presented this new game theory at a conference convened at the Institute for Advanced Studies on March 18, 2002, co-chaired by Stephen G. Post and Jeffrey P. Schloss.

12. Denis Alexander, *Rebuilding the Matrix* (Oxford: Lion Publishing, 2001), 363.

13. Edward O. Wilson, *On Human Nature* (Cambridge: Harvard University Press, 1978), 155.

14. Richard D. Alexander, *The Biology of Moral Systems* (New York: Aldine de Gruyter, 1987), 3.

15. David P. Barash, *Revolutionary Biology: The New, Gene-Centered View of Life* (New Brunswick, N.J.: Transaction, 2001), 38.

16. Ibid., 42.

17. Ibid., 46.

18. Ibid., 84, 85, 96.

19. Ibid., 106.

20. Ibid., 108.

21. Ibid., 114–115.

22. Ibid., 118.

23. Ibid., 133.

24. Ibid., 129.

25. Matt Ridley, *The Origins of Virtue: Human Instincts and the Evolution of Cooperation* (New York: Penguin, 1996), 63.

26. Ibid., 65.

27. Ibid., 84.

28. Ibid., 139.

29. Ibid., 145.

30. Ibid., 188.

31. Ibid., 193.

32. Melvin Konner, *The Tangled Wing: Biological Constraints on the Human Spirit* (New York: Henry Holt, 1982).

33. Peter Kropotkin, *Mutual Aid: A Factor in Evolution* (Montreal, Canada: Black Rose Books, 1989 [original 1903]), 300.

34. Elliott Sober and David Sloane Wilson, *Unto Others: The Evolution and Psychology of Unselfish Behavior* (Cambridge: Harvard University Press, 1998).

35. Ibid., 6.

36. Ibid., 8.

37. Ibid., 8–9.

38. Ibid., 330, 337.

39. Charles Darwin, *The Descent of Man* (Amherst, N.Y.: Prometheus Books, 1998 [original 1874]), 137.

40. Axelrod, *Evolution of Cooperation.*

41. Lee Dugatkin, *Cheating Monkeys and Citizen Bees: The Nature of Cooperation in Animals and Humans* (New York: Free Press, 1999).

42. Robert H. Frank, *Passions Within Reason: The Strategic Role of the Emotions* (New York: Norton, 1988), 221.

43. Paul R. Ehrlich, *Human Natures: Genes, Cultures, and the Human Prospect* (Washington, D.C.: Island Press, 2000), 312.

Notes to Chapter 6

1. Holmes Rolston III, *Genes, Genesis, and God: Values and Their Origins in Natural and Human History* (Cambridge: Cambridge University Press, 1999), 52, 82.

2. Steven Pinker, *How the Mind Works* (New York: W. W. Norton, 1997), 401.

3. Charles Darwin, *The Descent of Man* (Amherst, N.Y.: Prometheus Books, 1998 [original 1874, as revised from 1871 first edition], 109.

4. Ibid., 114.

5. Ibid., 121.

6. Ibid., 126–127.

7. Ibid., 137.

8. Charles Darwin, *The Expression of the Emotions in Man and Animals* (Chicago: University of Chicago Press, 1965 [original 1872]), 213.

9. William McDougall, *An Introduction to Social Psychology*, 23rd ed. (New York: Barnes & Noble, 1961 [original 1908]), 66–67.

10. Edward O. Wilson, *On Human Nature* (Cambridge: Harvard University Press, 1978), 155–156.

11. Ibid., 157.

12. Elliott Sober and David Sloane Wilson, *Unto Others: The Evolution and Psychology of Unselfish Behavior* (Cambridge: Harvard University Press, 1998), 302.

13. Ibid., 304.

14. Ibid., 304.

15. Ibid., 326.

16. See Robert Trivers, "Parent-Offspring Conflict," *American Zoologist*, vol. 14 (1974): 249–264.

17. Robert Wright, *The Moral Animal* (New York: Vintage Books, 1995), 57.

18. Stephen G. Post, "History, Infanticide, and Imperiled Newborns," *The Hastings Center Report*, vol. 18, no. 4 (August-September 1988): 18–23.

19. This is not a new idea. See, for example, in addition to McDougall's *Introduction to Social Psychology*, A. Sutherland, *The Origin and Growth of the Moral Instinct* (New York: Longmans, Green, & Co., 1998). See also Strachan Donnelley, "Natural Responsibilities: Philosophy, Biology, and Ethics in Ernst Mayr and Hans Jonas," *The Hastings Center Report*, vol. 42, no. 4 (July-August 2002): 36–43.

20. Hans Jonas, *The Imperative of Responsibility: In Search of an Ethics for the Technological Age* (Chicago: University of Chicago Press, 1984), 130–135.

21. John Boswell, *The Kindness of Strangers: The Abandonment of Children in Western Europe from Late Antiquity to the Renaissance* (New York: Pantheon Press, 1988), 37–38.

22. Steven Mithen, *The Prehistory of the Mind: A Search for the Origins of Art, Religion and Science* (London: Phoenix, 1996), 172.

23. Mithen, *Prehistory of the Mind*, 177. See also Michael Balter, "New Light on the Oldest Art," *Science*, vol. 283, no.12 (February 1999): 920–922.

24. Mithen, *Prehistory of the Mind*, 185.

25. Ibid., 200.

26. Elliott Sober, "Kindness and Cruelty in Evolution," in Richard J. Davidson and Anne Harrington, eds., *Visions of Compassion: Western Scientists and Tibetan Buddhists Examine Human Nature* (New York: Oxford University Press, 2002), 63.

Notes to Chapter 7

1. John Fiske, *Through Nature to God* (Boston: Houghton, Mifflin & Co., 1899), 120–121.

2. Susan Allport, *A Natural History of Parenting* (New York: Three Rivers Press, 1997), 20, 25–27.

3. Thomas Lewis, Fari Amini, and Richard Lannon, *A General Theory of Love* (New York: Random House, 2000), 25.

4. Ibid., 201.

5. See Jeffrey Moussaieff Masson and Susan McCarthy, *When Elephants Weep: The Emotional Lives of Animals* (New York: Delacorte Press, 1995).

6. Reinhold Niebuhr, *An Interpretation of Christian Ethics* (New York: Meridian Books, 1956 [original 1934]), 183.

7. Ibid., 185.

8. Ibid., 189.

9. Julian of Norwich, *Revelations of Divine Love,* trans. Elizabeth Spearing (London: Penguin, 1998 [original c. 1388]), 143.

10. C. S. Lewis, *The Four Loves* (San Diego: Harcourt Brace & Co., 1960), 31–32.

11. Sally McFague, "Mother God," in *Motherhood: Experience, Institutions, Theology,* ed. Ann Carr and Elizabeth Schussler Fiorenza (Edinburgh: T & T Clark, 1989), 139–140.

12. Ivone Gebara, "The Mother Superior and Spiritual Motherhood: From Intuition to Institution," in *Motherhood: Experience, Institutions, Theology,* 48.

13. Ibid., 139.

14. Dorothy Emmet, *The Nature of Metaphysical Thinking* (New York: St. Martin's Press, 1945), 6.

15. Daniel Day Williams, *The Spirit and the Forms of Love* (New York: Harper & Row, 1968), 122.

16. Paul Ramsey, *Fabricated Man: The Ethics of Genetic Control* (New Haven: Yale University Press, 1970), 38.

17. Emil Brunner, *The Divine Imperative,* trans. Olive Wyon (Philadelphia: Westminster Press, 1937), 346.

18. John Burnaby, *Amor Dei: A Study in the Religion of St. Augustine* (London: Hodder & Stoughton, 1938), 302.

19. Marie-Theres Wacker, "God as Mother? On the Meaning of a Biblical God-symbol for Feminist Theology," in *Motherhood: Experience, Institutions, Theology,* 105.

20. See Abraham J. Heschel, *God in Search of Man* (New York: Free Press, 1959).

21. Abraham Heschel, *The Prophets,* 2 vols. (New York: Harper & Row, 1962).

22. Willard Gaylin, *Adam and Eve and Pinnochio: On Being and Becoming Human* (New York: Viking, 1990), 214.

23. Louis Colin, C.S.S.R., *Love One Another,* trans. Fergus Murphy (Westminster, Md.: The Newman Press, 1960), 175–176, 178.

24. Paul Tillich, *Love, Power, and Justice* (New York: Oxford University Press, 1954), 116.

25. Ibid., 119.

26. Henri J. M. Nouwen, *The Return of the Prodigal Son: A Meditation on Fathers, Brothers, and Sons* (New York: Doubleday, 1992), 90.

27. Ibid., 94, 102.

28. Daniel Mark Epstein, *Love's Compass: A Natural History of the Heart* (Reading, Mass.: Addison-Wesley, 1990), 16, 41.

29. Feodor Dostoyevsky, *The Adolescent (or A Raw Youth)*, in *The Gospel in Dostoyevsky: Selections From His Works,* ed. the Hutterite Brethren (Ulster, N.Y.: Plough Publishing House, 1988), 219.

30. Ibid., 229.

31. Ibid., 221.

Notes to Chapter 8

1. George F. R. Ellis, "Kenosis as a Unifying Theme for Life and Cosmology," in John Polkinghorne, ed., *The Work of Love: Creation as Kenosis* (Grand Rapids, Mich.: Wm. B. Eerdmans Press, 2001), 108.

2. David P. Barash, *Revolutionary Biology: The New, Gene-Centered View of Life* (New Brunswick, N.J.: Transaction Publications, 2001), 139.

3. Ibid., 149.

4. Patty Anglin with Joe Musser, *Acres of Hope: The Miraculous Story of One Family's Gift of Love to Children without Hope* (Uhrichsville, Ohio: Promise Press, 1999).

5. William Werpehowski, "The Vocation of Parenthood: A Response to Stephen Post," *Journal of Religion Ethics,* vol. 25, no. 1 (spring 1997): 178.

6. Judith S. Modell, *Kinship with Strangers: Adoption and Interpretations of Kinship in American Culture* (Berkeley: University of California Press, 1994), 2.

7. Mary Watkins and Susan Fisher, *Talking with Young Children about Adoption* (New Haven: Yale University Press, 1993).

8. Marshall D. Schecter and Doris Bertocci, "The Meaning of the Search," in David M. Brodzinsky and Marshall D. Schecter, eds., *The Psychology of Adoption* (New York: Oxford University Press, 1990), 62–90.

9. D. Nelkin and M. S. Lindee, *The DNA Mystique: The Gene as a Cultural Icon* (New York: W. H. Freeman Press, 1995), 2.

10. Marque-Luisa Mirangoff, *The Social Costs of Genetic Welfare* (New Brunswick, N.J.: Rutgers University Press, 1991), 24.

11. Werpehowski, "The Vocation of Parenthood," 177.

12. James M. Scott, *Adoption as Sons of God: An Exegetical Investigation into the Background of Huiothesia in the Pauline Corpus* (Tubingen, Germany: J.C.B. Mohr, 1992).

13. Kristen Elizabeth Gager, *Blood Ties and Fictive Ties: Adoption and Family Life in Early Modern France* (Princeton: Princeton University Press, 1996).

14. Jack Goody, *The Development of the Family and Marriage in Europe* (Cambridge: Cambridge University Press, 1983), 68–73, 93–96.

15. Harry Hendrick, *Child Welfare: England 1872–1989* (New York: Routledge, 1994).

16. See Jean Bethke Elshtan, "The Chosen Family," a review article in *The New Republic,* 14 and 21 September, 1998, 47.

17. Elizabeth McKeown, "Adopting Sources: A Response to Stephen Post," *Journal of Religious Ethics,* vol. 25, no. 1 (spring 1997): 169–175.

18. Ibid., 155–156.

19. Ibid., 271.

20. Ibid., 155, 178.

21. Elie Spitz, "Through Her I Too Shall Bear a Child: Birth Surrogates in Jewish Law," *Journal of Religious Ethics,* vol. 24, no. 1 (1996): 65–97.

22. Garth A. Hallett, *Priorities and Christian Ethics* (Cambridge: Cambridge University Press, 1998), 112.

23. Ibid., 94.

24. Ibid., 95.

25. Insofar as Hallett refers to my previous writings, suffice it to state that if my review of his past work in any way encouraged his recent attention to "priorities" in Christian ethics, then I am honored. See Stephen G. Post, Review of Garth A. Hallett, *"Christian Neighbor-Love: An Assessment of Six Rival Versions,"* in *Journal of Religious Ethics,* vol. 19 (1991): 196.

26. Rodney Clapp, *Families at the Crossroads: Beyond Traditional and Modern Options* (Downers Grove, Ill.: InterVarsity Press, 1993), especially chap. 8, "No Christian Home Is a Haven."

27. See Henri J. M. Nouwen, *The Wounded Healer* (London: Darton, Longman & Todd, 1997 [original 1979]).

28. Louis Janssens, "Norms and Priorities in a Love Ethics," *Louvain Studies,* vol. 6 (1977): 212.

29. See Stephen J. Pope, "The Order of Love, and Recent Catholic Ethics: A Constructive Proposal," *Theological Studies,* vol. 52, no. 2 (June 1991): 255–288. This is an excellent criticism of recent Catholic ethics for failing to attend sufficiently to the ordering of love. Pope develops his recovery of Aquinas through the insights of certain aspects of evolutionary biology. See also Pope's *Love, Human Nature and Christian Ethics* (Washington, D. C.: Georgetown University Press, 1994).

30. See Hallett, *Priorities and Christian Ethics,* 7.

31. Josiah Royce, *The Problem of Christianity* (New York: Macmillan, 1931), 172–213.

32. Augustine, *On Christian Doctrine,* in *Augustine, Great Books of the Western World,* trans. J. F. Shaw (Chicago: Encyclopedia Britannica, 1952 [original c. 427]), 632.

33. Thomas Aquinas, *Summa Theologiae,* trans. D. J. Sullivan (Chicago: Encyclopedia Britannica Press, 1952), II–II, q. 26, a. 10.

34. Hallett, *Priorities and Christian Ethics,* 55–60, 62, 72.

35. Henry Sidgwick, *The Methods of Ethics* (Indianapolis: Hackett Publishing, 1981 [original 1906]), 241.

36. Adam Smith, *The Theory of Moral Sentiments,* ed. D. D. Raphael and A. L. Macfie (Indianapolis: Liberty Classics, 1982 [original 1759]), 216.

37. Joseph Butler, "Fifteen Sermons," in *British Moralists 1650–1800,* ed. D. D. Raphael (Oxford: Clarendon Press, 1969 [original 1726]), 373–374.

38. Lawrence C. Becker, *Reciprocity* (London: Routledge & Kegan Paul, 1986), 392.

39. For a brief summary of historical material, see Jeffrey Blustein, *Parents and Children: The Ethics of the Family* (New York: Oxford, 1982), sec. 1.

40. Thomas Aquinas, *Summa Theologiae,* q. 26, a. 8.

41. Max Scheler, *Ressentiment,* ed. Lewis A. Coser, trans. William W. Holdheim (New York: Free Press [original 1915], 1961), 96, 115–116 (emphasis in original).

42. Dorothy L. Sayers, *The Mind of the Maker* (San Francisco: HarperCollins, 1987 [original 1941], 25.

43. Ibid., 23.

44. Sir John Eccles and Daniel N. Robinson, *The Wonder of Being Human* (New York: Free Press, 1984).

Notes to Chapter 9

1. Huston Smith, *Religions of Man* (New York: Harper & Row, 1957), 24.

2. See John Polkinghorne, ed., *The Work of Love: Creation as Kenosis* (Grand Rapids, Mich.: Wm. B. Eerdmans, 2001).

3. Charles S. Pierce, "Evolutionary Love," in *Philosophical Writings of Pierce,* ed. Justice Buchler (New York: Dover, 1955 [original 1893]), 361.

4. Mark Caldwell, *A Short History of Rudeness: Manners Morals, and Misbehavior in Modern America* (New York: Picador, 1999).

5. See Edwin Post, *Truly Emily Post* (New York: Funk & Wagnalls, 1961).

6. See Lawrence Edwards Carter, Sr., ed., *Walking Integrity: Benjamin Elijah Mays, Mentor to Martin Luther King, Jr.* (Macon, Ga.: Mercer University Press, 1998).

7. See Dietrich Bonhoeffer, *Letters and Papers from Prison,* ed. Eberhard Bethge (New York: Macmillan, 1972 [original 1953]).

8. Teilhard de Chardin, *The Phenomenon of Man* (New York: Harper & Row, 1975 [1955 original]), 267.

9. William James, *The Varieties of Religious Experience* (New York: Penguin Books, 1982 [original 1902]), 272–274, 278–279.

10. Owsei Temkin, *Hippocrates in a World of Pagans and Christians* (Baltimore: Johns Hopkins University Press, 1991), 32.

11. John Ferguson, *Moral Values in the Ancient World* (New York: Barnes & Noble, 1959).

12. Darrel W. Amundsen, "Medical Ethics, History of: Europe, 2. Christianity," in Warren T. Reich, ed., *The Encyclopedia of Bioethics,* 2nd ed., 5 vols. (New York: Macmillan Reference, 1995), vol. 3, 1519.

13. Henry Sigerist, *Civilization and Disease* (Ithaca: Cornell University Press, 1943).

14. Edith Wyschogrod, *Saints and Postmodernism: Revisioning Moral Philosophy* (Chicago: University of Chicago Press, 1990).

15. Christine D. Pohl, *Making Room: Recovering Hospitality as a Christian Tradition* (Grand Rapids, Mich.: Wm. B. Eerdmans, 1999), 16.

16. W. E. H. Lecky, *History of European Morals from Augustus to Charlemagne* (New York: George Braziller, 1955).

17. See Paul Ramsey, *Basic Christian Ethics* (Chicago: University of Chicago Press, 1978), chap. 7.

18. See Gene Outka, *Agape: An Ethical Analysis* (New Haven: Yale University Press, 1972); see also Outka's "Equality and Individuality: Thoughts on Two Themes in Kierkegaard," *Journal of Religious Ethics,* vol. 10, no. 2 (fall 1982): 171–203.

19. Cited in Gilbert C. Meilaender, *Friendship: A Study in Theological Ethics* (Notre Dame, Ind.: University of Notre Dame Press, 1981), 1.

20. See Paul Ramsey, *War and the Christian Conscience* (Durham, N.C.: Duke University Press, 1961).

21. See Marjorie Hewitt Suchocki, *The Fall to Violence: Original Sin in Relational Theology* (New York: Continuum, 1999).

22. Cited in Jules Abel, *The Rockefeller Billions* (New York: Macmillan, 1968), 279.

23. Millard Fuller, with Diane Scott, *No More Shacks: The Daring Vision of Habitat for Humanity* (Waco, Tex.: Word Books, 1986).

24. Søren Kierkegaard, *Works of Love,* trans. Howard and Edna Hong (New York: Harper & Row, 1962 [original 1847]).

25. See John Wall, Don Browning, William J. Doherty, and Stephen G. Post, eds., *Marriage, Health, and the Professions* (Grand Rapids, Mich.: Wm. B. Eerdmans, 2002).

26. See Robert Hazo, *The Idea of Love* (New York: Praeger, 1967).

27. Kierkegaard, *Works of Love.*

28. Alan Soble, *The Structure of Love* (New Haven: Yale University Press, 1990), 15.

29. See Irving Singer, *The Nature of Love: The Modern World* (Chicago: University of Chicago Press, 1987).

30. Willard Gaylin, *Rediscovering Love* (New York: Penguin, 1986), 11.

31. Steven Seidman, *Embattled Eros: Sexual Politics and Ethics in Contemporary America* (New York: Routledge, 1992), 10.

32. Paul R. Fleischman, *The Healing Zone: Religious Issues in Psychotherapy* (New York: Paragon House, 1989), 173.

33. Ibid., 174.

34. Rollo May, *Love and Will* (New York: Dell Publishing, 1969), 42.

35. Catherine A. MacKinnon, *Toward a Feminist Theory of the State* (Cambridge: Harvard University Press, 1989).

36. C. S. Lewis, *Mere Christianity* (New York: Macmillan, 1952), 78, 81.

37. Denis de Rougemont, *Love in the Western World*, trans. M. Belgion (New York: Harper & Row, 1956), 41.

38. Stephen G. Post, *More Lasting Unions: Christianity, the Family, and Society* (Grand Rapids, Mich.: Wm. B. Eerdmans, 2000).

39. See Stephen G. Post, *Christian Love and Self-Denial: An Historical and Normative Study of Jonathan Edwards, Samuel Hopkins, and American Theological Ethics* (Lanham, Md.: University Press of America, 1987), 12.

40. See Stephen G. Post, "Disinterested Benevolence: An American Debate over the Nature of Christian Love," *Journal of Religious Ethics,* vol. 14, no. 2 (fall 1986): 356–368.

41. Ann Douglas, *The Feminization of American Culture* (New York: Avon Books, 1977), 50.

42. Valerie Saiving, "The Human Situation: A Feminine View," in *Womanspirit Rising,* ed. Carol Christ and Judith Plaskow (New York: Harper, 1979), 25–42.

43. See Stephen G. Post, *A Theory of Agape: On the Meaning of Christian Love* (Lewisburg, Pa.: Bucknell University Press, 1990).

44. See Stephen G. Post, "Communion and True Self Love," *Journal of Religious Ethics,* vol. 16, no. 2 (fall 1988): 345–362.

45. Jaroslav Pelikan, *Jesus Throughout the Centuries: His Place in the History of Culture* (New York: Harper & Row, 1985), 95, 100.

46. Victor Paul Furnish, *Theology and Ethics in Paul* (Nashville: Abingdon, 1968), 163.

47. I have written more fully on these topics in the following: "Disinterested Benevolence: An American Debate Over the Nature of Christian Love"; *A Theory of Agape*; and *Spheres of Love: Toward a New Ethics of the Family* (Dallas, Tex.: Southern Methodist University Press, 1995).

48. Anne Sebba, *Mother Teresa: Beyond the Image* (New York: Doubleday, 1997), xi.

Notes to Chapter 10

1. M. E. McCullough, R.A. Emmons, and J. Tsang, "The Grateful Disposition: A Conceptual and Empirical Topography," *Journal of Personality and Social Psychology,* vol. 82, no. 1 (2002): 112–127; see also M. E. McCullough, S. Kirkpatrick, R. A. Emmons, and D. Larson, "Is Gratitutde a Moral Affect?" *Psychological Bulletin,* vol. 127 (2001): 249–266.

2. Robert A. Emmons and Joanna Hill, *Words of Gratitude For Mind, Body, and Soul* (Philadelphia, Pa.: Templeton Foundation Press, 2001), 15.

3. D. C. McClelland, "Some Reflections on the two Psychologies of Love." *Journal of Personality*, vol. 54 (1986): 334–53.

4. See J. Levin, "A Prolegomenon to an Epidemiology of Love: Theory, Measurement, and Health Outcomes," *Journal of Social and Clinical Psychology*, Vol. 19, no. 1 (2000): 117–136.

5. W. R. Miller, "Rediscovering Fire: Small Interventions, Large Effects," *Psychology of Addictive Behaviors*, vol. 14, no., 1 (2000): 6–18.

6. See Thomas Lewis, Fari Amini, and Richard Lannon, *A General Theory of Love* (New York: Random House, 2000); see also John Bowlby, *Attachment* (New York: Basic Books, 1969).

7. G. T. Reker, "Personal Meaning, Optimism, and Choice: Existential Predictors of Depression in Community and Institutional Elderly," *The Gerontologist*, vol. 37, no. 6 (1997): 709–716.

8. M. Van Willigen, "Differential Benefits of Volunteering Across the Life Course," *Journals of Gerontology* Series B., vol. 55B, no. 5 (2000): S308–S318.

9. D. Oman, C. E. Thoresen, K. McMahon, "Volunteerism and Mortality among the Community-Dwelling Elderly," *Journal of Health Psychology*, vol. 4, no. 3 (1999): 301–316.

10. B. J. Fisher, "Successful Aging, Life Satisfaction, and Generativity in Later Life," *International Journal of Aging and Human Development*, vol. 41, no. 2 (1995): 239–250.

INDEX